SpringerBriefs in Applied Sciences and Technology

PoliMI SpringerBriefs

For further volumes:
http://www.springer.com/series/11159
http://www.polimi.it

Saeed Eftekhar Azam

Online Damage Detection in Structural Systems

Applications of Proper Orthogonal Decomposition, and Kalman and Particle Filters

POLITECNICO
DI MILANO

Springer

Saeed Eftekhar Azam
Road, Housing and Urban Development
 Research Center
Tehran
Iran

ISSN 2282-2577   ISSN 2282-2585 (electronic)
ISBN 978-3-319-02558-2   ISBN 978-3-319-02559-9 (eBook)
DOI 10.1007/978-3-319-02559-9
Springer Cham Heidelberg New York Dordrecht London

Library of Congress Control Number: 2013956335

Printed on acid-free paper

Springer is part of Springer Science+Business Media (www.springer.com)

Foreword

Monitoring the health of structures and infrastructures exposed to aging or extreme loadings is nowadays recognized as a societal need. The pervasive use of miniaturized sensors, recently developed through microelectronics-driven technological processes, has forced people to look for smart monitoring strategies tailored to handle the large amount of data provided by densely deployed sensor networks. Moreover, as each health monitoring procedure relies upon a theoretical/numerical model of the considered structure, the more accurate the model the more powerful the monitoring scheme; such increased accuracy also entails additional monitoring burden.

If a structure undergoes a damaging process reducing its load-carrying capacity, the health monitoring procedure should be able to identify the damage itself in terms of location and amplitude. It is then necessary to filter out the possible noise terms and provide meaningful information from the structural response. Due to the presence of damage, robust procedures able to deal with a nonlinear system evolution are obviously to be envisioned.

The two topics discussed above, i.e., the size of the model to be handled and the nonlinearities in its evolution law, might be difficult to manage simultaneously in a common frame. It may happen that the filtering algorithm, which is supposed to compare the responses of the real structure and of a fictitious, linear-comparison one (featuring no damage evolution in a predefined time window), provides estimations affected by drifts or biases, sometimes also diverging. It may also happen that by increasing the size of the numerical model, e.g., due to a required finer space discretization in case of numerical (e.g., finite element) procedures supplying the model itself, the aforementioned bias and divergence issues get amplified.

It is also worth noting that structural health monitoring systems should be able to provide results in real-time or, at least, close to such target, so that warnings can be provided as soon as critical conditions are approached during the life cycle of a structure.

The research activity reported in this book moved from all the aforementioned critical aspects, with the aim of providing a robust, accurate, and easy to implement methodology for the health monitoring of civil structures and infrastructures, possibly suffering damage inception and growth. Two main challenging topics are specifically dealt with: the derivative-free filtering of the response of nonlinear systems; a time-varying, reduced-order modeling able to self-adapt to a changing system dynamics. As for the former issue, results are known to be not satisfactory if one does not properly account for the statistics of noise terms and structural state, and for the nonlinear evolution of the last ones. Here, the author shows that a wise combination of Kalman and particle filtering can indeed provide a very efficient (in terms of computational costs) and robust (in terms of avoidance of output divergence) framework. As for the latter issue, a snapshot-driven proper orthogonal decomposition methodology is known to work well in case of time-evolving linear systems; on the other hand, it is still disputed whether proper orthogonal decomposition can be adopted for a nonlinear time evolution of the system, linked, e.g., to damage growth. Here, the author shows that a further exploitation of Kalman filtering can provide, if governed by a partial observation of the system, a very efficient way to continuously tune the reduced-order model, thereby avoiding time-consuming re-training stages suggested by others in the past.

This book thus introduces a novel, hybrid approach to damage identification and health monitoring of structural systems. As such, it has been written mainly focusing on the theoretical and implementational aspects of the approach, partially leaving experimental validations aside. In my opinion, readers can find in it all the details necessary to adapt the methodology to many, if not all the real-life situations to be practically envisioned.

Stefano Mariani

Preface

The aim of this monograph is to present the key ingredients of a still-in-progress research discipline within the structural engineering realm, namely online damage detection. The material of the text offers detailed explanations on recursive Bayesian filters (e.g., Kalman filters, particle filters), proper orthogonal decomposition methods (POD) (e.g., singular value decomposition, principal component analysis), and a combination thereof, i.e., a synergy of reduced order modeling and recursive Bayesian filtering. Illustrations accompanied by the theoretical description allow the reader to intuitively comprehend the notions. Therefore, this book can serve as a tutorial for scientists and engineers who want to apply and implement proper orthogonal decomposition and/or Bayesian filters to a specific problem.

Throughout the book, the focus of the numerical examples is on structural systems. The techniques presented in this research monograph are well established in fields like automatic control, statistics, etc. However, they are rather new to civil and structural engineers; hence, the algorithms are presented in enough details so that the reader can easily implement them on any structural state-space model. At first, the ease of implementation has been the main concern; however, the author believes that the way the main notions are analyzed makes this book an inspiration for conducting further research and development of these methods.

The objective of the study presented in this monograph is to develop techniques for vibration-based non-destructive damage identification of the structures. In fact, the major emphasis is on the development of quick and robust recursive damage detection algorithms in order to facilitate the task of online, real-time continuous monitoring of civil structures, such as, e.g., residential buildings, bridges, and other similar structures. This goal can be accomplished only through mixing different disciplines of science and technology, including automatic control, applied mathematics, and structural engineering.

It should be emphasized that though Bayesian filters have been extensively studied in the automatic control field, their applications in structural engineering are yet to be investigated. The applications of extended Kalman filter (EKF), sigma-point Kalman filter (SPKF), and particle filter (PF) to simplified and low-dimensional models are suggested in the existing literature; nevertheless, to the best of my knowledge, applying the extended Kalman-particle filter (EK-PF) has never been reported when dealing with a structural engineering problem.

The algorithms for all the Bayesian filters used in this book are derived using the same notation; this can allow the reader to easily understand the similarities and ideas behind each one of them. Their performances dealing with different identification tasks are scrutinized in detail, and the reason for their success and failure in each case is highlighted.

It is perceived that as the number of the degrees-of-freedom increases, the adopted methods in the literature lose their accuracy in system identification, and thus in damage detection process. This problem is created due to the high dimension of the parameter space, i.e., by so-called curse of dimensionality. To manage this issue, in this study I make recourse to reduced order modeling of the systems. The aforementioned task is accomplished by using the proper orthogonal decomposition. Before using POD-based models in the Bayesian filters, the performance of such methodology is thoroughly investigated to ensure accuracy, speed-up, and robustness when different sources of excitation shake the structures.

The major contribution of the present research is the development of a recursive stochastic algorithm by a synergy of dual estimation concept, POD-based order reduction, and a subspace update. The proposed methodology takes advantage of Bayesian filters (like EKF and EK-PF) for dual estimation of state and parameters of a reduced order model of a time-varying system. A Kalman filter is employed within each iteration period to update the subspace spanned by the POMs of the structure. The efficiency and effectiveness of the algorithm are verified via pseudo-experimental tests conducted on multi-storey shear buildings. It will be shown that the procedure successfully identifies the state, the model parameters (i.e., the components of the reduced stiffness matrix of the structure) and relevant proper orthogonal modes (POMs) of the reduced model. Unbiased estimates furnished by the algorithm permit the health monitoring of the structure.

By reading this monograph, one could learn how the family of Kalman filters and particle filters are connected; compare their performances when dealing with a structural dynamics problem; see through detailed examples why and when they fail; figure out which filter can better fit a certain problem; and know how to tune the parameters of the filters. Moreover, the way the filters are presented renders the task of implementing more complicated filters easy and even developing ad hoc filters for structural engineering possible. Concerning reduced order modeling, possible limitations caused by POD-based reduced models are shown via numerous graphs and tables. The nature and extent of the inaccuracies caused by abridging the full mathematical model of the structures are carefully studied and analyzed. Finally, the use of such reduced models in the Bayesian filters is studied for the case in which the model can change (sustain damage) and also when it is a priori known that the model remains undamaged.

To follow the contents of this monograph, the reader is expected to have a background in statistics and calculus, and to be familiar with linear algebra and fundamentals of signal processing.

The material covered in this book is derived from the doctoral dissertation of the author, which was submitted to the scientific faculty of the doctoral program at the Department of Structural Engineering of the Politecnico di Milano. The author sincerely acknowledges the role of his supervisor Prof. Stefano Mariani in shaping this monograph; his encouragements, friendliness, rational way of thoughts, mindful directions, and scientific attitude had been the main elements without which this work would never have been in its present state. The author wants to thank all his friends in the department and outside it, whose presence made those years so memorable.

Saeed Eftekhar Azam

Contents

Chapter 1
Introduction

Abstract In the current Chapter, the fact that a significant portion of the existing civil structures and infrastructures in the developed and industrialized nations was constructed at the early period of twentieth century is discussed. It is also expressed that a notable part of the existing structures have been subject to deteriorations. Moreover, the need to develop damage identification techniques for vibration based non-destructive damage identification of the structures is briefly debated. Then, the major emphasis of the monograph and the type of the target structures is explained. In the end, the major disciplines covered inside the book are highlighted.

1.1 Background and Motivation

A significant portion of civil structures and infrastructures was constructed at the early period of twentieth century in the developed and industrialized nations; consequently, they have been subject to deteriorations. To illustrate this issue, over 50 % of the bridges were built in the U.S.A prior to 1940 (Stallings et al. 2000); moreover, over 42 % of all the aforementioned bridges are structurally deficient as reported by Klaiber et al. 1987. In Canada, over 40 % of the present functional bridges were built prior to 1970 and majority of these Canadian bridges demand prompt rehabilitation, strengthening or replacement (ISIS Canada 2007). The Canadian Construction Association estimated nearly 900 billion US dollars as the cost to rehabilitate global infrastructures (ISIS Canada 2007). In the next years, it takes a great deal of budget to rehabilitate the global infrastructure which highlights the significance of developing reliable and cost effective methods for the investments required for rehabilitation. Moreover, in seismically active zones, the deterioration due to degradation in the structures may be combined with the damage due to extreme seismic actions.

In recent years, civil engineering community has globally focused their attention on structural health monitoring with the purpose to identify the damage

S. Eftekhar Azam, *Online Damage Detection in Structural Systems*,
PoliMI SpringerBriefs, DOI: 10.1007/978-3-319-02559-9_1,
© The Author(s) 2014

occurred in civil structures at the earliest possible stage, and to estimate the remaining lifetime of the structures. Structural damage caused by corrosion leads to degradation of the mechanical properties of the affected components; therefore, it changes the response of the structure as well. Moreover, the failure of the structural components such as shear walls, bracings and connections clearly changes the mathematical system which is defined to mimic the behavior of the structure. Hence, the goal of structural health monitoring can be perceived by structural system identification. The system corresponding to healthy state should be primarily identified; moreover possible changes which occur in the system with respect to the healthy state of structures are indications of structural damage in next planned system identifications. This task is realized within the frames of non-destructive vibration-based damage identification either by direct identification of the system or an alternative indirect scheme. Moreover, several dynamic charac-teristics of the system are identified, and possible variations in their value are employed to update the system. Instances of former methods include dual esti-mation of states and parameters of the structure via Bayesian inference techniques (Chatzi et al. 2010), while latter methods utilize modal properties of the structure to detect the damage (Moaveni et al. 2010).

To prevent the possible casualties and losses caused by sudden collapse of the structure, timely detection of the structural damage is essential.

The collapse of the bridge on Minneapolis I-35 W highway is one of the recent structural catastrophe. The steel truss bridge, constructed in 1967, collapsed during rush hour which led to dozens of causalities on August 1, 2007 (French et al. 2011). Beyond philanthropic issues, the economic impact of the collapse has been substantial: road-user costs due to the unavailability of the river crossing imposed a financial burden of $220,000 US dollars per day (Xie and Levinson 2011). These statistics highlight the economic significance of the civil infrastructure; and therefore substantiate the demand to monitor their safety: St. Anthony Falls Bridge on the I35 W, constructed to replace the collapsed steel truss bridge, includes over 500 instruments to monitor the structural behavior (French et al. 2011). To detect the damage at the earliest possible stage, long-term monitoring systems are required to process the data sensed by these instruments.

1.2 Objectives and Scope

The objective of the study presented in this monograph is to develop damage identification techniques for vibration based non-destructive damage identification of the structures. In fact, the major emphasis is on the development of quick and robust recursive damage detection algorithms in order to facilitate the task of online real-time continuous monitoring of civil structures, such as e.g. residential buildings, bridges and other similar structures. To accomplish this end, four Bayesian filters, namely the extended Kalman filter (EKF), the sigma-point Kalman filter (SPKF), the particle filter (PF) and a hybrid extended Kalman particle filter

(EK-PF) are adopted to identify the structural system. To avoid shadowing effects of the structural system, performance of the filters is benchmarked by dual estimation of state and parameters of a single degrees-of-freedom structure featuring nonlinear behaviors namely: an exponential softening and a bilinear (linear-softening, linear plastic and linear hardening) constitutive laws are examined. It will be observed that the EK-PF outperforms all the other filters studied in this research. It should be emphasized that though Bayesian filters have been extensively studied in the automatic control field, their applications in structural engineering is yet to be investigated. The applications of EKF, SPKF, and PF to simplified and low dimensional models are suggested in the existing literature; nevertheless, to the best of our knowledge, applying EK-PF has never been reported when dealing with a structural engineering problem. After the performance of the filters is benchmarked when working with a single degree-of-freedom system, multi degrees-of-freedom structures are handled. Consequently, EKF for its computational efficiency and EK-PF for its excellent performance working with single degree-of-freedom systems are adopted. It will be indicated that the performance of EKF and EK-PF is identical when engaging with a two degrees-of-freedom system; nevertheless, moving to three and four degrees-of-freedom structures, the EK-PF outperforms the EKF in terms of the bias in the estimation. It is perceived that as the number of the degrees-of-freedom increases, the adopted methods lose their accuracy in system identification and thus in damage detection process. This problem is created due to the high dimension of the parameter space, i.e. by the so-called curse of dimensionality. To manage this issue in this study, we make recourse to reduced order modeling of the systems. Regarding the model order reduction technique, a method based on the proper orthogonal decomposition (POD) is adopted. Such method utilizes POD to define a subspace in which the main dynamic evolution of the system occurs; the vectors that span the POD subspace are called proper orthogonal modes (POMs). Once such a subspace is attained, a projection method onto the POD subspace is employed to reduce the order of the set of governing equations of the system; subsequently the speed of calculations is increased. In addition to the expediting the calculations, another striking property of the so-called POMs is that they are sensitive to changes in the system parameters; thus in this study this property is exploited to identify the damage in the structure.

The major novel contribution of this monograph is to develop a recursive stochastic algorithm by a synergy of dual estimation concept, POD-based order reduction and subspace update. The proposed methodology takes advantage of Bayesian filters (e.g. EKF and EK-PF) for dual estimation of state and parameters of a reduced order model of a time-varying system. A Kalman filter is employed within each iteration period to update the subspace spanned by the POMs of the structure. The efficiency and effectiveness of the algorithm is verified via pseudo-experimental tests conducted on a ten-storey shear building. It will be indicated that the procedure successfully identifies the state, the model parameters (i.e. the components of the reduced stiffness matrix of the structure) and relevant POMs of the reduced model. Unbiased estimates furnished by the algorithm permits the health monitoring of the structure.

1.3 Organization of the Content

The present research is categorized into three major topics in this monograph, namely: (a) online and model-based system identification of dynamic systems; (b) model order reduction of dynamic systems; and (c) reduced order model identification of dynamic systems. The content of the monograph is derived from the PhD dissertation of its author which has been presented to the faculty of the doctoral course of the department of structural engineering at the Technical University of Milan (Eftekhar Azam 2012).

In the Chap. 2, the first research topic is extensively examined. Dual estimation of state and parameters of structural state space models is considered; moreover, the EKF, SPKF, PF and EK-PF are employed for parameter identification and state estimation. First, the performance of the filters is benchmarked by applying a single degree-of-freedom nonlinear system; subsequently, application of the filters to multi degrees-of-freedom systems is considered. Therefore, a multi storey shear building is assessed. Limitations for applicability of this approach in the identification of e.g. the stiffness matrix of multi storey structures are highlighted. It is concluded that due to bias in the estimates, these approaches are not suitable for system identification of shear building structures with more than three storeys.

Model order reduction of multi storey buildings is presented in the Chap. 3. Proper orthogonal decomposition is employed to extract the minimal subspace which features the dominant characteristics of the structure, via information contained in the response of the structure itself. The subspace discovered by POD is obtained by mathematical manipulation of the samples of the response of the structure (gathered in the so-called snapshot matrix), thus it can be load dependent. In case the external excitation is formerly known, load dependency of the reduced model will not be a problem; nevertheless in case of seismic excitations, such condition is not always true. To address this issue and build the snapshot matrix, the samples are selected from the response of a case-study structure to the El Centro accelerogram; furthermore, the obtained reduced model is subsequently employed to simulate the response of the case-study structure to the Friuli and the Kobe earthquake records. It is observed that POD-based reduced models are robust to changes in input seismic load. Afterwards, efficiency of the method in expediting the calculations, with high level of fidelity, is numerically examined.

Chapter 4, investigates the statistical properties of residual errors induced by POD-based reduced order modeling. Such errors enter the state space equations of the reduced systems in terms of system evolution and observation noise. A fundamental assumption made by recursive Bayesian filters, as exploited in this study, is the whiteness of the aforementioned noises. In this chapter, null hypothesis of the whiteness of the noise signals is tested by utilizing the Bartlett's whiteness test.

It is indicated that, no matter what the number of POMs retained in the analysis is, the null hypothesis of the whiteness is constantly to be rejected. Nevertheless, the spectral power of the embedded periodic signals decreases rapidly by increasing the number of POMs. The speed-up gained by incorporating POD-based reduced models into Kalman observer of linear time invariant systems is stated in this chapter as well. Chapter 5 tackles the major objective of this research: the dual estimation of the reduced order model, and update of POMs of the structure to provide damage detection in structural system. It is revealed that the first POM of the structure is quite sensitive to the intensity and location of the damage: a reduced model, featuring even a single POM retains enough insights to be used in developing damage detection algorithms. The proposed procedure exhibits an acceptable performance when applied to pseudo-experimental tests. It is indicated that the algorithm estimates the state, model parameters and relevant POMs of the reduced model of a ten storey shear building, featuring convergence to the true values of parameters and the POMs employed to create the pseudo test.

Final chapter of the monograph is allocated to the conclusions and suggestions for future study. It is noteworthy that this monograph proposes a novel methodology based on recursive Bayesian inference of a reduced order model of the structure. Accuracy and power of the proposed approach has been tested in this work through pseudo-experimental analysis. Online and real-time detection of the damage in the civil structural systems is a field which is yet to be investigated. It is suggested to utilize other existing Bayesian filtering techniques for the objective of the online real-time damage detection. Since this study does not provide experimental verification of the proposed methodology; hence it is recommended as a future research project.

References

Chatzi EN, Smyth AW, Masri SF (2010) Experimental application of on-line parametric identification for nonlinear hysteretic systems with model uncertainty. Struct Saf 32:326–337

Eftekhar Azam S (2012) Dual estimation and reduced order modelling of damaging structures. Doctoral dissertation, Italy

French CE, Hedegaard B, Shield CK, Stolarski H (2011) I35 W collapse, rebuild, and structural health monitoring—challenges associated with structural health monitoring of bridge systems. AIP Conf Proc 1335:9–30

ISIS Canada (2007) Reinforcing concrete structures with fibre reinforced polymers. ISIS Canada Design Manual, University of Manitoba, Winnipeg

Klaiber FW, Dunker KF, Wipf TJ, Sanders WW (1987) Methods of strengthening existing highway bridges. Transp Res Rec 1380:1–6

Moaveni B, He X, Conte JP, Restrepo JI (2010) Damage identification study of a seven-story full-scale building slice tested on the UCSD-NEES shake table. Struct Saf 32:347–356

Stallings JM, Tedesco JW, El-Mihilmy M, McCauley M (2000) Field performance of FRP bridge repairs. J Bridge Eng 5:107–113

Xie F, Levinson D (2011) Evaluating the effects of the I-35 W bridge collapse on road-users in the twin cities metropolitan region. Transp Plan Technol 34:691–703

Chapter 2
Recursive Bayesian Estimation
of Partially Observed Dynamic Systems

Abstract In the current Chapter, recursive Bayesian inference of partially observed dynamical systems is reviewed. As a tool for structural system identification, nonlinear Bayesian filters are applied to dual estimation problem of linear and nonlinear dynamical systems. In so doing, dual estimation of state and parameters of structural state space models is considered; EKF, SPKF, PF and EK-PF are used for parameter identification and state estimation. Dealing with a SDOF structure, it is shown that the hybrid EK-PF filter is able to furnish a reasonable estimation of parameters of nonlinear constitutive models. Assessment of SDOF systems is followed by identification of multi storey buildings. In this regard, performances of the EK-PF and EKF algorithms are compared, and it is concluded that they are nearly the same, and by an increase in the number of storeys of the building, both of the algorithms fail to provide an unbiased estimate of the parameters (stiffness of the storeys). Therefore, they are not reliable tools to monitor state and parameters of multi storey systems.

2.1 Introduction

Recursive inference of the dynamics of a system through noisy observations is normally pursued within a Bayesian framework. As a result, if there is a priori information available on probability distribution of observable quantities of the system and there is a correlation between observable and hidden quantities of the system, Bayes probability concept is employed to estimate probability distribution of the hidden state variables. Extensive variety of applications are exploited by using such approach namely: in econometrics to estimate volatility in the market (Ishihara and Omori 2012; Yang and Lee 2011; Miazhynskaia et al. 2006), for a review on the literature see (Creal 2012). In field of robotics, this approach is applied to develop behaviors for robots (Lazkano et al. 2007), system identification of the robots (Ting et al. 2011), and their localization (Zhou and Sakane 2007). In biology, this approach is employed for molecular characterization of diseases

S. Eftekhar Azam, *Online Damage Detection in Structural Systems*,
PoliMI SpringerBriefs, DOI: 10.1007/978-3-319-02559-9_2,
© The Author(s) 2014

(Alvarado Mora et al. 2011), finding linkage in DNA (Allen and Darwiche 2008; Biedermann and Taroni 2012) and for characterization of genomic data (Caron et al. 2012). In image processing, this approach is used to diagnose diseases from medical images (Mitra et al. 2005), for image segmentation (Adelino and Ferreira da Silva 2009), and for image retrieval (Duan et al. 2005). Moreover, this approach is employed in the following fields such as: object tracking and radars (Jay et al. 2003; Velarde et al. 2008; White et al. 2009); in speech enhancement (Saleh and Niranjan 2001; Yahya et al. 2010); in mechanical characterization and parameter identification of materials (Corigliano and Mariani 2004, 2001a, b; Bittanti et al. 1984), mechanical system identification (Mariani and Ghisi 2007; Mariani and Corigliano 2005) and many other fields which are not mentioned for the sake of brevity. The aforementioned instances are just a few fields of application of Bayesian inference schemes; their diversity proves the versatility of such approach in problem solving processes.

Estimation of state and parameters of a structural system are simultaneously are dealt with in a recursive fashion in this chapter of the monograph. As new observations become available, the information concerning the current state of the system, which is attained through a model of the system, is updated based on the measured observation. This objective is perceived by utilizing four recursive Bayesian filters, namely: the extended Kalman filter (EKF), the sigma-point Kalman filter (S-PKF), the particle filter (PF), and a newly proposed hybrid extended Kalman particle filter (EKPF). Therefore, to avoid shadowing effects of high dimensional structures, a single degree-of-freedom system has primarily been considered. The performances of the filters are standardized to simultaneously estimate state and parameters of a nonlinear constitutive model of the system. After the performance of the filters working with a single degree-of-freedom structure has been verified, we move to the analysis of multi degree-of-freedom (DOF) structures. To accomplish this aim, a shear type of buildings has been considered. It should be emphasized that although Bayesian filters under the study have been adopted in the other fields such as automatic control, their application in the field of structural engineering demands further investigations. The author of this book has coauthored three articles on peer reviewed international journals on this topics (Eftekhar Azam and Mariani 2012; Eftekhar Azam et al. 2012a, b). The proceeding parts of this Chapter is classified as follows: in Sect. 2.2, the dual estimation concept for simultaneous estimation of state and parameters of a state-space model is reviewed. General frames of the recursive Bayesian inference techniques are discussed in Sect. 2.3; moreover, the Kalman filter, as the optimal filter of linear state-space models is devoted to Sect. 2.4. Approximate Bayesian filters for nonlinear systems are dealt with in Sect. 2.5; furthermore in Sect. 2.6 the numerical results concerning dual estimation of states and parameters of single DOF and multi DOFs structures are presented. The Chapter is eventually concluded in Sect. 2.7, where the limitations filters under the study are discussed together with our remedy to solve the issue when applied to simultaneous state and parameter estimation of high dimensional problems.

2.2 Dual Estimation of States and Parameters of Mechanical Systems

In this research, the emphasis is on civil structures. Therefore, mechanical systems whose dynamics is governed by the famous set of ordinary differential equations are addressed which governs evolution of their dynamic:

$$M\ddot{u} + D\dot{u} + R(u, t) = F(t) \tag{2.1}$$

where M is assigned as the mass matrix, D represents the damping matrix; $R(u, t)$ stands for possibly displacement dependent internal force, whereas $F(t)$ is designated as the loading vector; u, $\dot{u}$ and $\ddot{u}$ are the nodal displacements, velocities and accelerations, respectively. Since measurements are normally completed in discrete time, our attention is limited to a discrete time formulation, where it is assumed that a part of displacements or accelerations of the system are measured in evenly spaced time grids.

To embed the mathematical model into algorithms designed for recursive Bayesian inference, we represent the dynamics of the system in a state-space form; the details concerning the state-space representation of the mathematical model (2.1) are presented in the following Sections. Throughout the book, displacement, velocity and acceleration quantities of the response of the structure are assigned by the word 'state' and we intend to use 'parameters' which represent the coefficients of the internal force term (in linear elastic case, components of the stiffness matrix). The state vector z thus contains u, $\dot{u}$ and $\ddot{u}$, namely:

$$z_k = \begin{bmatrix} u_k \\ \dot{u}_k \\ \ddot{u}_k \end{bmatrix} \tag{2.2}$$

while parameter vector ϑ_k collects several unknown parameters of the system.

The state space representation of the system thus is expressed as:

$$z_k = f_k^z(z_{k-1}; \vartheta_{k-1}) + v_k^z \tag{2.3}$$

$$y_k = H_k^z z_k + w_k \tag{2.4}$$

where for any time interval $[t_{k-1} t_k]$, $f_k^z(.)$ is a function of the state z_{k-1} and parameters ϑ_{k-1} of the system, and evolves the state of the system z_{k-1} to obtain z_k. H_k^z quantifies the correlation between the state and the observable part of the system, at any given time instant; the name of Eq. (2.4), observation equation, originates from the aforementioned fact. v_k^z and w_k are the zero mean, uncorrelated Gaussian processes with covariance matrices V^z and W, respectively. Generally, observation equation may take any form; however, in the present study, it is reasonably assumed that observation process consists of a part of the state vector, namely displacements and/or accelerations of several representative points. As a result, the observation equation can be expressed as a sum of a linear mapping of the state through a Boolean matrix $(\boldsymbol{H}_k^z)$ and an additive, uncorrelated Gaussian noise stemming from uncertainty of measurement sensor.

In this study, the major mission of Bayesian filters, beyond estimating hidden part of the state vector, will be the calibration of system model parameters in an online method. At each time interval $[t_{k-1}t_k]$, on the basis of the information contained in the latest observation $\boldsymbol{y}_k$, the algorithms update former knowledge of the parameter $\boldsymbol{\vartheta}_{k-1}$ to yield $\boldsymbol{\vartheta}_k$. To accomplish this objective, dual estimation of states and parameters are considered; hence, the parameter vector $\boldsymbol{\vartheta}_k$ is increased by defining the state vector (Mariani and Corigliano 2005):

$$\boldsymbol{x}_k = \begin{bmatrix} \boldsymbol{z}_k \\ \boldsymbol{\vartheta}_k \end{bmatrix}. \tag{2.5}$$

In addition to the conventional form of state-space equation, which is composed of evolution and observation equations, dual estimation is pursued via an extra vectorial equation governing the evolution of the parameters over time according to:

$$\boldsymbol{\vartheta}_k = \boldsymbol{\vartheta}_{k-1} + \boldsymbol{v}_k^{\vartheta}. \tag{2.6}$$

The intuitive idea underlying this extra equation is to allow the unknown parameters of the system to change over time, starting from an initial guess and hopefully converge on an unbiased estimate. The possibility of variation to parameters is provided by white Gaussian fictitious noise $\boldsymbol{v}_k^{\vartheta}$, added to parameter evolution. Moreover, the intensity of such a noise should be tuned in order to obtain an unbiased and converging estimate for the parameters (Bittanti and Savaresi 2000). The state-space equation governing evolution of the increased state thus is expressed as:

$$\boldsymbol{x}_k = \boldsymbol{f}_k(\boldsymbol{x}_{k-1}) + \boldsymbol{v}_k \tag{2.7}$$

$$\boldsymbol{y}_k = \boldsymbol{H}_k\boldsymbol{x}_k + \boldsymbol{w}_k \tag{2.8}$$

where $\boldsymbol{f}_k(.)$, maps the extended state vector $\boldsymbol{x}_k$ over time; therefore, it features both Eqs. (2.3) and (2.6) in one unique equation.

2.3 Recursive Bayesian Inference

The inference problem can be considered as recursively estimating the expected value $E[\boldsymbol{x}_k|\boldsymbol{y}_{1:k}]$ of the state of the system, conditioned on the observations. If the initial probability density function (PDF) $p(\boldsymbol{x}_0|\boldsymbol{y}_0) = p(\boldsymbol{x}_0)$ of the state vector is known, the problem consists in estimating $p(\boldsymbol{x}_k|\boldsymbol{y}_{1:k})$, assuming that the conditional PDF $p(\boldsymbol{x}_{k-1}|\boldsymbol{y}_{1:k-1})$ is available. The problem can be broken down into in two stages of prediction and update. As far as the prediction stage, the Chapman–Kolmogorov equation furnishes the a-priori estimate of the state PDF at t_k (Arulampalam et al. 2002):

$$p(x_k|y_{1:k-1}) = \int p(x_k|x_{k-1})\,p(x_{k-1}|y_{1:k-1})dx_{k-1}. \tag{2.9}$$

In the updating stage, as soon as the new observation y_k becomes available, Bayes rule is profited to apply correction on the PDF of the state (Cadini et al. 2009):

$$p(x_k|y_{1:k}) = \varsigma\, p(y_k|x_k)p(x_k|y_{1:k-1}) \tag{2.10}$$

where ς is stands for a normalizing constant which depends on the likelihood function of the observation process. The Eqs. (2.9) and (2.10) collectively forge the basis for any Bayesian recursive inference scheme. The analytical solution of the integral in (2.9) is not possible except for a limited category of problems, namely systems which are formulated by linear state space equations and disturbed by uncorrelated white Gaussian noises (Eftekhar Azam et al. 2012a). In case of a general nonlinear problem, one should make recourse to approximate solutions, either by approximating the nonlinear evolution equations via linearization (Corigliano and Mariani 2004) or via discrete approximate representation of the PDF of the state vector (Mariani and Ghisi 2007; Doucet and Johansen 2009; Doucet and Johansen 2009). In the next Section, the major features of the analytical solution available for linear Gaussian state space model are examined, and is followed by the Sect. 2.5 which handles approximate solutions for nonlinear state-space models.

2.4 Linear Dynamic State Space Equations: Optimal Closed Form Estimator

As addressed in the preceding section, recursive Bayesian estimation of linear Gaussian state-space models can be calculated analytically. A linear discrete state-space model is considered which can be obtained by substituting the arbitrary evolution equation $f_k(.)$ in Eqs. (2.7) and (2.8) by a linear operator F_k. Therefore, the state-space equations of such a system are expressed as:

$$x_k = F_k x_{k-1} + v_k, \tag{2.11}$$

$$y_k = H_k x_k + w_k. \tag{2.12}$$

If the primary probability distribution of the state is Gaussian, it is straightforward to display that a linear operator does not change the Gaussian PDF over time (Kalman 1960). That is, in the Chapman–Kolmogorov integral at any arbitrary time instant t_k, the functional form of both integrands is a priori known; *moreover, $p(x_{k-1}|y_{1:k-1})$* is constantly a Gaussian probability density function, and $p(x_k|x_{k-1})$ is by definition a Gaussian function as well. As a result, the integral can

Table 2.1 Kalman Filter
algorithm

- Initialization at time t_0:
$$\widehat{x}_0 = \mathbb{E}[x_0]$$
$$P_0 = \mathbb{E}\left[(x_0 - \widehat{x}_0)(x_0 - \widehat{x}_0)^\mathrm{T}\right]$$
- At time t_k, for $k = 1,\ldots,N_t$:
- Prediction stage:
    1. Evolution of state and prediction of covariance
    $$x_k^- = F_k x_{k-1}$$
    $$P_k^- = F_k P_{k-1} F_k^\mathrm{T} + V$$
- Update stage:
    1. Calculation of Kalman gain:
    $$G_k = P_k^- H_k^\mathrm{T}\left(H_k P_k^- H_k^\mathrm{T} + W\right)^{-1}$$
    2. Improve predictions using latest observation:
    $$\widehat{x}_k = x_k^- + G_k\left(y_k - H_k x_k^-\right)$$
    $$P_k = P_k^- - G_k H_k P_k^-$$

be calculated analytically. Kalman introduced a well-known filter which is the optimal estimator for linear systems with uncorrelated Gaussian noise in his seminal study (Kalman 1960); the filter provides an online estimation of first and second order statistics of a state space model, and it includes a prediction stage which is simply an evolution of state over time. In the updating stage, by computing the Kalman gain G_k, the filter enhances the predicted values furnished in previous stage. Readers are referred to Table 2.1 for a detailed description and algorithmic implementation of the Kalman filter (KF).

2.4.1 The Kalman Filter

In many real life problems, neither the dynamics of the system takes a linear form nor the uncertainties of transition equation which may be regarded as Gaussian distributions. Even if the initial distribution of the uncertainties could be assumed Gaussian, a nonlinear state-space model would change the distribution over time (Mariani and Ghisi 2007). Therefore, an optimal closed form solution will not be available for a general nonlinear problem (Doucet and Johansen 2009).

In a mechanical system, the source of nonlinearity might be the material response to loading (Corigliano and Mariani 2001a, b; Corigliano 1993); however, even if the material behavior is linear, dual estimation of states and parameter will result in a bilinear (nonlinear) state space model (Ljung 1999). We illustrate this issue via an intuitive example by considering the following linear state space model:

$$z_k = az_{k-1} + b + v_k^z \tag{2.13}$$

$$y_k = Hz_k + w_k \qquad (2.14)$$

where z_k and y_k denote the state and the observation of the system at a given time instant t_k; a and b represent the linear transition for the state in a given time interval $[t_{k-1}t_k]$, while H links the hidden state z_k to the observation process. v_k^z and w_k denote the zero mean white Gaussian processes which quantify evolution and measurement inaccuracies, respectively. In case one is only interested in estimating the state of the system z_k, we already know the Kalman filter furnishes the optimal estimation; however, let us imagine one is also interested in an online estimation of the parameters of the state space model. For the sake of simplicity, we assume that only parameter a is of interest. As aforementioned, the trick in dual estimation framework is to collect the unknown parameter a into the extended state vector x_k and try to track the dynamics of such system via recursive Bayesian inference algorithms. It is noteworthy that even though parameter a is stationary by definition, the parameter is allowed to vary in the formulation of dual estimation. In this regard, a transition equation governing evolution of the parameter is introduced:

$$a_k = a_{k-1} + v_k^a. \qquad (2.15)$$

Equation (2.15), together with (2.13) and (2.14), constitute the required state-space model for dual estimation of states and parameters. The augmented state vector x_k thus becomes $x_k = [z_k \quad a_k]^T$, where $x_k(1) = z_k$ and $x_k(2) = a_k$; consequently Eqs. (2.13–2.15) become:

$$x_k(1) = x_{k-1}(2)x_{k-1}(1) + b \qquad (2.16)$$

$$x_k(2) = x_{k-1}(2) + v_k^a \qquad (2.17)$$

$$y_k = Hx_k(1) + w_k \qquad (2.18)$$

or, in matrix form:

$$\begin{bmatrix} x_k(1) \\ x_k(2) \end{bmatrix} = \begin{bmatrix} x_{k-1}(2)x_{k-1}(1) \\ x_{k-1}(2) \end{bmatrix} + \begin{bmatrix} v_k^z \\ v_k^a \end{bmatrix} + \begin{bmatrix} b \\ 0 \end{bmatrix} \qquad (2.19)$$

$$y_k = [H \quad 0] \begin{bmatrix} x_k(1) \\ x_k(2) \end{bmatrix} + w_k. \qquad (2.20)$$

It is evident that Eq. (2.19) is a nonlinear mapping over the given time interval $[t_{k-1}t_k]$. The aforementioned fact, together with consideration that many real life problems are nonlinear, substantiates the need for Bayesian inference algorithms targeting general nonlinear, non-Gaussian problems. The following Section is devoted to review the approximate solutions available in the literature to deal recursive Bayesian estimation of nonlinear state-space models.

2.5 Nonlinear Dynamic State Space Equations: Approximate Bayesian Estimators

Most of the problems in the real problems, as well as all the problems related to the identification of the parameters of the systems by use of dual estimation concept lead to nonlinear state-space models. Hence, developing nonlinear versions of the KF seemed inevitable from the very beginning. Next subchapter reviews the main concepts behind the extension of the KF to the nonlinear problems.

2.5.1 The Extended Kalman Filter

A simple remedy to deal with nonlinear state-space models is through an extension of the Kalman filter, where for each time instant t_k, the nonlinear state mapping $f_k(x_{k-1})$ is linearized by a first order truncation of a Taylor series expansion around x_{k-1}. To accomplish this goal, the Jacobian of the evolution equation is used as a surrogate for linear transition matrices in order to update covariance (Gelb 1974); subsequently, the Kalman gain is used to update state statistics. This procedure is the extension of the Kalman filter for nonlinear state space models; thus its name extended Kalman filter (EKF). The extended Kalman filter assumes the prior $p(x_{k-1}|y_{1:k-1})$ to be Gaussian; however, even if initially Gaussian, a nonlinear mapping changes its probability distribution. Moreover, a severely nonlinear mapping of state might change the probability distribution into a tailed or a bimodal distribution (Adelino and Ferreira da Silva 2009; Van der Merwe 2004) and causes bias in the estimates furnished by the EKF. In addition, the approximation of the state mapping via its Jacobian is not accurate enough in several cases; for instance, it does not consider the stochastic nature of the state vector, and the effect of the neglected terms may become considerable. As a consequence, the approximation may lead to an inconsistent estimation of the covariance; hence, a bias or divergence may occur in estimation of the state (Julier and Uhlmann 1997). For a detailed description of EKF algorithm see Table 2.2, where $\nabla_x f_k(x)|_{x=x_{k-1}}$ denotes the Jacobian of $f_k(x)$ at $x = x_{k-1}$. To Alleviate the aforementioned issues posed by highly nonlinear models the first remedy has been the development of the sigma-point Kalman filter which will be discussed in the next subsection.

2.5.2 The Sigma-Point Kalman Filter

In case of severely nonlinear systems, the successive linearization approach may be inaccurate (Mariani 2009b). For certain problems, it may be practically difficult to adopt: in case of a non holonomic material behavior, to calculate the Jacobian,

Table 2.2 Extended Kalman filter algorithm

- Initialization at time t_0:
$$\widehat{x}_0 = \mathbb{E}[x_0]$$
$$P_0 = \mathbb{E}\left[(x_0 - \widehat{x}_0)(x_0 - \widehat{x}_0)^{\mathrm{T}}\right]$$
- At time t_k, for $k = 1, \ldots, N_t$:
- Prediction stage:
    1. Computing process model Jacobian:
    $$F_k = \nabla_x f_k(x)|_{x = x_{k-1}}$$
    2. Evolution of state and prediction of covariance:
    $$x_k^- = F_k x_{k-1}$$
    $$P_k^- = F_k P_{k-1} F_k^T + V$$
- Update stage:
    1. Calculation of Kalman gain:
    $$G_k = P_k^- H_k^{\mathrm{T}}\left(H_k P_k^- H_k^{\mathrm{T}} + W\right)^{-1}$$
    2. Improve predictions using latest observation:
    $$\widehat{x}_k = x_k^- + G_k\left(y_k - H_k x_k^-\right)$$
    $$P_k = P_k^- - G_k H_k P_k^-$$

one should know if the current state of the system proceeds toward loading or unloading (Mariani and Ghisi 2007). The difficulty in estimation of the Jacobian and also its inadequate accuracy has led to development of a category of derivative-free filters, called sigma-point Kalman filters, SPKF (Julier et al. 1995, 2000). The basic idea behind these filters is that it is easier to approximate a probability distribution compared to a nonlinear state-space model. A SPKF uses a deterministic set of quadrature points, called sigma-points, to handle the Chapman–Kolmogorov integral (Ito and Xiong 2000). This set of deterministic points can be used since a-prior distribution of the state is assumed to have a Gaussian functional form for all the time instants. Hence, it is possible to approximate it through a set of deterministic points which are parameterized through the mean and covariance of the state vector. The distribution of the state vector, a multivariate Gaussian probability distribution, at time t_{k-1} reads:

$$p(x_{k-1}|y_{1:k-1}) = \frac{1}{((2\pi)^n|P_{k-1}|)^{1/2}} \exp[-\frac{1}{2}(x_{k-1} - \widehat{x}_{k-1})^T P_{k-1}^{-1}(x_{k-1} - \widehat{x}_{k-1})]$$

(2.21)

where $\widehat{x}_{k-1}$ and P_{k-1} are the associated mean vector and covariance matrix of the state vector, respectively.

Once it is established that the a priori distribution of the state vector is a known Gaussian one, the Chapman–Kolmogorov integral can be recast as a Gaussian integral of the form $\int_{\mathbb{R}^n} \mu(x)\omega(x)dx$, where $\mu(.)$ is an arbitrary probability distribution, whereas $\omega(.)$ denotes the a priori probability distribution of state vector. Hence (2.9) becomes (Ito and Xiong 2000):

$$\int \mu(x_{k-1}) \frac{1}{((2\pi)^n |P_{k-1}|)^{1/2}} \exp[-\frac{1}{2}(x_{k-1} - \widehat{x}_{k-1})^T P_{k-1}^{-1}(x_{k-1} - \widehat{x}_{k-1})]dx_{k-1}$$

$$(2.22)$$

where $\omega(.)$ is an arbitrary function of state vector. To numerically handle the Gaussian integral in (2.22), a discrete representation of (2.21) is necessary as done by a set of points which feature the same statistics of the original Gaussian distribution (Ito and Xiong 2000):

$$\chi_j = \begin{cases} \sqrt{n + \rho}e_j, & 1 \le j \le n \\ -\sqrt{n + \rho}e_{j-n}, & n+1 \le j \le 2n \\ \mathbf{0}, & 2n+1 \end{cases} \qquad (2.23)$$

and

$$\omega(\chi_j) = \begin{cases} \dfrac{2\rho}{2(n+\rho)} & j = 2n+1 \\ \dfrac{1}{2(n+\rho)} & 1 \le j \le 2n \end{cases} \qquad (2.24)$$

where $\rho > 0$ is a constant and e_k is the kth unit vector in $\mathbb{R}^n$. Julier and co-workers (Julier et al. 1995) proposed their S-PKF based on a quadrature rule which, for scalar functions, is identical to the Gauss-Hermit quadrature rule (Ito and Xiong 2000):

$$\int_{\mathbb{R}^n} \mu(x)\omega(x)dx \approx \sum_{i=1}^{2n+1} \mu(\chi_i)\omega(\chi_i). \qquad (2.25)$$

The $2n + 1$ quadrature points are the minimal number of points necessary to preserve the first and the second moments of a multivariate normal distribution (Julier et al. 1995). One can assume $\omega(\chi_i)$ as quadrature weights, which in this case are constant in all time instants, while the quadrature points vary over time on the basis of the information contained in the covariance of the state, at $t = t_k$ the set of sigma-points are:

$$\chi_{k,j} = \begin{cases} \widehat{x}_{k-1} & j = 2n+1 \\ \widehat{x}_{k-1} + \psi\sqrt{P_{k-1,j}} & 1 \le j \le n \\ \widehat{x}_{k-1} - \psi\sqrt{P_{k-1,(j-n)}} & n+1 \le j \le 2n \end{cases} \qquad (2.26)$$

where $\widehat{x}_{k-1}$ denotes the expected value of the state and $\sqrt{P_{k-1,j}}$ stands for jth column of square root of its associated covariance at $t = t_{k-1}$. This scheme outperforms the extended Kalman filter (Mariani and Ghisi 2007); for a detailed description of SPKF algorithm, see Table 2.3.

In Table 2.3, ω^{*j} and ω^j are weights adopted in the merging stage at the end of the time step, to build mean and covariance of the current state. ψ instead denotes, a time invariant scaling factor which renders possible capturing local effects of nonlinear functions. To enhance the performance, the scaling factor ψ should be

Table 2.3 Sigma-Point
Kalman filter algorithm

- Initialization at time t_0:
$$\widehat{x}_0 = \mathbb{E}[x_0]$$
$$P_0 = \mathbb{E}\left[(x_0 - \widehat{x}_0)(x_0 - \widehat{x}_0)^T\right]$$
- At time t_k, for $k = 1, \ldots, N_t$:
- Prediction stage:
   1. Deploying sigma-points:
$$\chi_{k,j}^- = \begin{cases} \widehat{x}_{k-1} & j = 2n + 1 \\ \widehat{x}_{k-1} + \psi\sqrt{P_{k-1,j}} & 1 \leq j \leq n \\ \widehat{x}_{k-1} - \psi\sqrt{P_{k-1,(j-n)}} & n + 1 \leq j \leq 2n \end{cases}$$
   2. Evolution of the sigma points:
$$\chi_{k,j} = f_k\left(\chi_{k,j}^-\right)$$
   3. Estimation of the statistics:
$$x_k^- = \sum_{j=1}^{2n+1} \omega^j \chi_{k,j}$$
$$P_k^- = R_k + V$$
where
$$R_k = \sum_{j=1}^{2n+1} \omega^{*j}\left(\chi_{k,j} - x_k^-\right)\left(\chi_{k,j} - x_k^-\right)^T$$
- Update stage:
   1. Calculation of Kalman gain:
$$G_k = P_k^- H_k^T\left(H_k P_k^- H_k^T + W\right)^{-1}$$
   2. Improve predictions using latest observation:
$$\widehat{x}_k = x_k^- + G_k\left(y_k - H_k x_k^-\right)$$
$$P_k = P_k^- - G_k H_k P_k^-$$

carefully calibrated to allow appropriate capturing of local nonlinearity effects, by tuning the distances of each sigma-point from the mean of a priori distribution of the variable. In the SPKF, the square root $\sqrt{P_{k-1}}$ is calculated by a Choleski factorization. The subscript j refers to the jth column of the Choleski factor of the covariance.

The SPKF approach, similarly to the EKF, is based on the assumption that at each time instant, the a priori distribution of the state is Gaussian. To deal with a general class of nonlinear models, the particle filter approach has been developed by the academic society, the next subsection is devoted to highlight the main notions of it.

2.5.3 The Particle Filter

To deal with more general problems, it is a common practice to make recourse to Sequential Monte Carlo methods (Doucet and Johansen 2009) to handle the Chapman–Kolmogorov integral by numerical approximations. Sequential

Monte-Carlo methods make no explicit assumptions concerning the form of the posterior density $p(x_{0:k}|y_{1:k})$. These methods approximate the Chapman–Kolmogorov integrals in (2.9) through finite sums, adopting a sequential importance sampling on an adaptive stochastic grid. Within this frame, the particle filter implements an optimal recursive Bayesian estimation by recursively approximating the complete posterior state density. A set of N_P weighted particles $x_k^{(i)}$, drawn from the posterior distribution $p(x_{0:k}|y_{1:k})$, is used to map the integrals. To accomplish this objective, the main trick is to represent the posteriori PDF via Dirac delta functions pond at discrete sample points, namely the so-called particles. Without the loss of generality, one can write (Cadini et al. 2009):

$$p(x_{0:k}|y_{1:k}) = \int p(\varepsilon_{0:k}|y_{1:k})\, \delta(\varepsilon_{0:k} - x_{0:k})d\varepsilon_k \qquad (2.27)$$

where $\delta(.)$ denotes the Dirac function.

Assuming the true posterior $p(x_{1:k}|y_{1:k})$ is known and can be sampled, an estimated of (2.27) is given by:

$$p(x_{0:k}|y_{1:k}) \approx \frac{1}{N_P}\sum_{i=1}^{N_P} \delta(x_{0:k} - x_{0:k}^i) \qquad (2.28)$$

where x_k^i are a set of random samples drawn from true posterior $p(x_{0:k}|y_{1:k})$. In practice, it is impossible to efficiently sample from the true posterior; a remedy is built by making recourse to the importance sampling, i.e. to sample state sequences from an arbitrarily chosen distribution $\pi(x_{0:k}|y_{1:k})$ called importance function. An unbiased estimate of $p(x_{0:k}|y_{1:k})$ can then still be made as (Cadini et al. 2009):

$$
\begin{aligned}
p(x_{0:k}|y_{1:k}) &= \int \pi(\varepsilon_{0:k}|y_{1:k})\frac{p(\varepsilon_k|y_{1:k})}{\pi(\varepsilon_{0:k}|y_{1:k})}\,\delta(\varepsilon_{0:k} - x_{0:k})d\varepsilon_k \\
&\approx \frac{1}{N_s}\sum_{i=1}^{N_s} \omega_k^{*i}\delta(x_{0:k} - x_{0:k}^i)
\end{aligned}
\qquad (2.29)
$$

where

$$\omega_k^{*i} = \frac{p(y_{1:k}|x_{0:k}^i)p(x_{0:k}^i)}{p(y_{1:k})\pi(x_{0:k}^i|y_{1:k})} \qquad (2.30)$$

is the importance weight associated to the state process x_k^i. In practice, these weights are difficult to calculate, due to the need of evaluating the integral to normalize constant $p(y_{1:k})$. Instead, the following weights are used (Gordon et al. 1993):

$$\omega_k^i = \frac{p(y_{1:k}|x_{0:k}^i)p(x_{0:k}^i)}{\pi(x_k^i|y_{1:k})} \qquad (2.31)$$

which are subsequently normalized according to:

$$\tilde{\omega}_k^i = \frac{\omega_k^i}{\sum_{j=1}^{N_s} \omega_k^j}. \tag{2.32}$$

Thus, estimate of the posterior distribution reads:

$$p(\boldsymbol{x}_{0:k}|\boldsymbol{y}_{1:k}) \approx \sum_{i=1}^{N_s} \tilde{\omega}_k^i \delta\left(\boldsymbol{x}_{0:k} - \boldsymbol{x}_{0:k}^i\right). \tag{2.33}$$

If the current state of the importance sampling function does not depend on future observations, i.e. if the importance sampling function satisfies the following condition (Van der Merwe 2004):

$$\pi(\boldsymbol{x}_{1:k}|\boldsymbol{y}_{1:k}) = \pi(\boldsymbol{x}_1|\boldsymbol{y}_1) \prod_{j=1}^{k} \pi(\boldsymbol{x}_j|\boldsymbol{x}_{1:j-1}, \boldsymbol{y}_{1:j})$$

$$= \pi(\boldsymbol{x}_k|\boldsymbol{x}_{1:k-1}, \boldsymbol{y}_{1:k}) \pi(\boldsymbol{x}_{1:k-1}|\boldsymbol{y}_{1:k-1}) \tag{2.34}$$

and if states can be considered as a Markov process, through the assumption that the observations are conditionally independent, given the states we obtain (Van der Merwe 2004):

$$p(\boldsymbol{x}_{0:k}) = p(\boldsymbol{x}_{0:k}) \prod_{j=1}^{k} p(\boldsymbol{x}_j|\boldsymbol{x}_{j-1}) \tag{2.35}$$

$$p(\boldsymbol{y}_{1:k}|\boldsymbol{x}_{0:k}) = \prod_{j=1}^{k} p(\boldsymbol{y}_j|\boldsymbol{x}_j) \tag{2.36}$$

thus by using Eqs. (2.34–2.36) in (2.30), the recursive formula for the update of importance weights becomes (Van der Merwe 2004):

$$\omega_k^i = \omega_{k-1}^i \frac{p(\boldsymbol{y}_k|\boldsymbol{x}_k^i)p(\boldsymbol{x}_k^i|\boldsymbol{x}_{k-1}^i)}{\pi(\boldsymbol{x}_k^i|\boldsymbol{x}_{0:k-1}^i, \boldsymbol{y}_{1k})}. \tag{2.37}$$

For filtering purposes, the estimation of the marginal probability density $p(\boldsymbol{x}_k|\boldsymbol{y}_k)$ of the full posterior is sufficient, that is, if $\pi(\boldsymbol{x}_k|\boldsymbol{x}_{1:k-1}, \boldsymbol{y}_{1:k})$ is substituted by $\pi(\boldsymbol{x}_k|\boldsymbol{x}_{k-1}, \boldsymbol{y}_k)$, the sampling proposal will only depend on $\boldsymbol{x}_{k-1}$ and $\boldsymbol{y}_k$ (Arulampalam et al. 2002). Consequently, the recursive formula for estimation and update of the non-normalized weights is expressed as (Arulampalam et al. 2002):

$$\omega_k^i = \omega_{k-1}^i \frac{p(\boldsymbol{y}_k|\boldsymbol{x}_k^i)p(\boldsymbol{x}_k^i|\boldsymbol{x}_{k-1}^i)}{\pi(\boldsymbol{x}_k^i|\boldsymbol{x}_{k-1}^i, \boldsymbol{y}_k)}. \tag{2.38}$$

The (2.38) provides a method to sequentially update the importance weights, given an appropriate choice of the proposal distribution $\pi(\boldsymbol{x}_k|\boldsymbol{x}_{k-1}, \boldsymbol{y}_k)$. Consequently, any expectations of the form $E[g(\boldsymbol{x}_k)] = \int g(\boldsymbol{x}_k)p(\boldsymbol{x}_{0:k}|\boldsymbol{y}_{1:k})d\boldsymbol{x}_k$, $g(.)$ being any given function can be approximated through $E[g(\boldsymbol{x}_k)] \approx \sum_{j=1}^{N_P} \omega_k^i g(\boldsymbol{x}_k^i)$.

In (Doucet 1997), it was shown that the proposal distribution $\pi(x_k|x_{k-1}, y_k)$ minimizes the variance of the importance weights, conditional on x_{k-1} and y_k. Nonetheless, the distribution $p(x_j|x_{j-1})$ (i.e. the transition prior) is the most popular choice for the proposal distribution. Although it results in a Monte-Carlo variation higher than that obtained using the optimal proposal $\pi(x_k|x_{k-1}, y_k)$, the importance weights are easily updated by simply evaluating the observation likelihood density $\pi(x_k|x_{k-1})$ for the sampled particle set, through (Cadini et al. 2009):

$$\omega_k^i = \omega_{k-1}^i p(y_k|x_k^i). \tag{2.39}$$

The variance of these importance weights increases stochastically over time (Doucet 1997); after a few time steps, one of the normalized importance weights tends to one, while the remaining weights tend to zero. To address this rapid degeneracy, a resampling stage may be used to eliminate samples with low importance weights, and duplicate samples with high importance weights. An intuitive explanation of particle filtering technique is expressed as: each sample x_k^i might be a solution of the problem, and its associated weight ω_k^i signifies its probability of being the correct solution. In the resampling stage, the particles with higher probability are duplicated and in turn, the ones with lower probability are discarded. Such an approach somehow permits the filter to condense the cloud of particles around the peak probability zone. An algorithm built in this method was primarily proposed by Gordon et al. (1993), and has been called in different names such as bootstrap filter, condensation algorithm etc.; for a detailed algorithmic specification see Table 2.4.

It is worth underlining that the update stage in the particle filter algorithm is conducted via evolution of particle weights based on the probability of occurrence of each particle conditioned on the latest observation. After such weight evolution, the resampling stage is prescribed to alleviate the degeneracy issue, where ensemble of the samples is refined to increase the population of the samples which are more likely and decrease the lower probability population. To this end, different algorithms were proposed in the literature, namely e.g. stratified, systematic, or residual resampling. Accounting for sampling quality and computational complexity, the systematic resampling scheme adopted turns out to be the most favorable one in this study (Hol et al. 2006). The resampling stage is performed by drawing a random sample ζ_j from the uniform distribution over (0,1]; afterwards, the Mth particle for which the value of the random number ζ_j is between values of the empirical cumulative distribution of particles at $M - 1$ and M is duplicated by resampling stage. Details of the systematic resampling (Kitagawa 1996) algorithm are shown in Table 2.5.

Since particle filter handles the current, the actual PDF of the state to draw particles in prediction stage, it can appropriately account for non-Gaussian densities. However, as the dimension of the state vector increases, computational costs associated with numerical integrations increase drastically. It is suggested, as a rough rule of thumb, not to apply particle filter to problems with dimension of state vector more than five (Li et al. 2004).

Table 2.4 Particle filter algorithm

- Initialization at time t_0:

$$\widehat{x}_0 = \mathbb{E}[x_0] \qquad\qquad x_0^{(i)} = \widehat{x}_0$$
$$P_0 = \mathbb{E}\left[(x_0 - \widehat{x}_0)(x_0 - \widehat{x}_0)^{\mathrm{T}}\right] \quad \omega_0^{(i)} = p(y_0|x_0) \quad i = 1,\ldots,N_P$$

- At time t_k, for $k = 1,\ldots,N_t$:
- • Prediction stage:
  1. Draw particles:

  $$x_k^{(i)} \sim p\left(x_k|x_{k-1}^{(i)}\right) i = 1,\ldots,N_P$$

- • Update stage:
  1. Evolve weights:

  $$\omega_k^{(i)} = \omega_{k-1}^{(i)} p\left(y_k|x_k^{(i)}\right) i = 1,\ldots,N_P$$

  2. Resampling, see Table 2.5.
  3. Compute expected value:

  $$\widehat{x}_k = \sum_{i=1}^{N_P} \omega_k^{(i)} x_k^{(i)}$$

Table 2.5 Systematic resampling algorithm

- At time t_k, for $j = 1,\ldots,N_P$:
  - • draw a random sample ζ_j from uniform distribution over $(0,1]$
  - • find M that satisfies:

  $$\sum_{i=1}^{M-1} \omega_k^{(i)} < \zeta_j \leq \sum_{i=1}^{M} \omega_k^{(i)}$$

  - • $x_k^{(j)} \leftarrow x_k^{(M)}$

The sampling distribution used in the generic particle filter can cause serious problems, since it is not the optimal one and conditioned on the latest observation. This fact leads to high computational costs, since the cloud of the samples fall far from the zones with high probability; therefore, many samples have to be drawn in order to make the algorithm to converge.

2.5.4 The Hybrid Extended Kalman Particle Filter

To alleviate the issues discussed in the previous subsection, our remedy is to keep using the same sampling distribution; however, after the samples are drawn, we improve the quality of the ensemble of the samples. Roughly speaking, once the samples are drawn, they are pushed by an extended Kalman filter toward the zones of higher probability in order to incorporate data from the latest observations into each sample.

The reason for exploiting the EKF instead of the SPKF, for enhancing the quality of sample ensemble, is twofold: first, the difficulty in tuning it in a way to have all the particles moved appropriately; second, the computational cost of the SPKF combined with particle filter can be significant, since both adopt numerical

Table 2.6 Hybrid extended Kalman particle filter algorithm

- Initialization at time t_0:
$$\widehat{x}_0 = \mathbb{E}[x_0] \, x_0^{(i)} = \widehat{x}_0$$
$$P_0 = \mathbb{E}\left[(x_0 - \widehat{x}_0)(x_0 - \widehat{x}_0)^{\mathrm{T}}\right] \omega_0^{(i)} = p(y_0|x_0) \, i = 1, \ldots, N_{\mathrm{P}}$$
- At time t_k, for $k = 1, \ldots, N_t$:
- Prediction stage:
  1. Draw particles:
  $$x_k^{(i)} \sim p\left(x_k | x_{k-1}^{(i)}\right) i = 1, \ldots, N_{\mathrm{P}}$$
  2. Push the particles toward the region of high probability through an EKF:
  $$P_k^{(i)-} = F_k P_{k-1}^{(i)} F_k^{\mathrm{T}} + V$$
  $$G_k^{(i)} = P_k^{(i)-} H_k^{\mathrm{T}} \left(H_k P_k^{(i)-} H_k^{\mathrm{T}} + W\right)^{-1}$$
  $$x_k^{(i)} = x_k^{(i)-} + G_k^{(i)} \left(y_k - H_k x_k^{(i)-}\right)$$
  $$P_k^{(i)} = P_k^{(i)-} - G_k^{(i)} H_k P_k^{(i)-} \quad i = 1, \ldots, N_{\mathrm{P}}$$
- Update stage:
  1. Evolve weights:
  $$\omega_k^{(i)} = \omega_{k-1}^{(i)} p\left(y_k | x_k^{(i)}\right) i = 1, \ldots, N_{\mathrm{P}}$$
  2. Resampling, see Table 2.5.
  3. Compute expected value or other required statistics:
  $$\widehat{x}_k = \sum_{i=1}^{N_{\mathrm{P}}} \omega_k^{(i)} x_k^{(i)}$$

approximations to handle the quadrature. That is, the EKF is combined with particle filter frames to update each particle based on the information contained in the latest observation, see Table 2.6.

In Table 2.6, F_k represents the current Jacobians of mappings $f_k(\blacksquare)$.

In what follows, we will assess the performance of the filters through numerical examples. In the absence of experimental data, for validation of the algorithms, we rely on pseudo experimental data, i.e. numerical data resulting from direct analysis contaminated by white Gaussian processes substitute noisy measurements of the observable part of the state vector.

2.6 Numerical Results for Dual Estimation of Single Degree and Multi Degrees of Freedom Dynamic Systems

To numerically solve the set of ordinary differential equations that govern the dynamics of the system, a Newmark explicit integration scheme has been adopted. According to (Hughes 2000), the time marching algorithm within the time step $[t_{k-1} t_k]$ can be partitioned into:

- predictor stage:

$$\tilde{u}_k = u_{k-1} + \Delta t \dot{u}_{k-1} + \Delta t^2 (\frac{1}{2} - \beta)\ddot{u}_{k-1} \tag{2.40}$$

$$\dot{\tilde{u}}_k = \dot{u}_{k-1} + \Delta t (1 - \gamma)\ddot{u}_{k-1}; \tag{2.41}$$

- explicit integrator:

$$\ddot{u}_k = M^{-1}(R_k - (D\dot{\tilde{u}}_k + K\tilde{u}_k)); \tag{2.42}$$

- corrector stage:

$$u_k = \tilde{u}_k + \Delta t^2 \beta \ddot{u}_k \tag{2.43}$$

$$\dot{u}_k = \dot{\tilde{u}}_k + \Delta t \gamma \ddot{u}_k \tag{2.44}$$

where $\Delta t = t_k - t_{k-1}$ denotes the time step size. To ensure numerical stability in the linear regime, Δt needs to be upper bounded by Bathe (1996):

$$\Delta t_{cr} = \frac{T_n}{\pi} \tag{2.45}$$

where T_n is the period associated with the highest oscillation frequency. Even if Δt_{cr} can be increased in the reduced model, since higher order oscillations are filtered out of the numerical solution, in what follows we are keeping Δt constant in all the simulations. Hence, the speedup reported is therefore to be mainly linked to the reduction of the number of handled DOFs.

In Corigliano and Mariani (2001b), it was shown that structural effects may play a prominent role in system identification. They typically lead to shadowing effects, arising when the sensitivity of measurable variables (like, e.g. displacements or velocities) to constitutive parameters becomes negligible or falls out of the measurement range (i.e. they become comparable to round-off errors). Such structural effects practically lead to multiple solutions of the inverse problem in terms of model parameters update (all difficult to distinguish in the noisy environment), and filters provide biased or divergent calibrations, see e.g. (Corigliano and Mariani 2001a, b, 2004). To solely benchmark performance of the filters, we primarily focus on dynamics of a single degree-of-freedom structure. Once the performances of the filters are benchmarked by analyses concerning a single degree-of-freedom, then we move to the multi degrees of freedom structures to study the applicability of these methods in more realistic scenarios.

2.6.1 Single Degree-of-Freedom Dynamic System

Since we are interested in benchmarking the extended Kalman particle filter (EK-PF) when compared to other Bayesian filters tested in this study (i.e. the EKF, the SPKF and the PF), the aforementioned structural effects are avoided by focusing on an undamped single DOF system constituted by a mass (or rigid block) connected to the reference frame through a spring, see Fig. 2.1. The equation of motion of the system is expressed as:

$$M\ddot{u} + R(u) = F(t) \tag{2.46}$$

where m is the block mass; $R(u)$ is the spring force; $F(t)$ is the external load, which evolves in time; u and $\ddot{u}$ are the displacement and acceleration of the block, respectively. The results can be easily extended to the damped case; in such situation, it is, however, important to have the system continuously to be (or permanently) excited, so as to avoid vibration amplitudes to progressively decrease in time, thereby loosing filter efficiency, see (Corigliano and Mariani 2004).

In this study, all the studied filters perform well for dual estimation of a linear SDOF structure; hence, the results are not discussed for the sake of brevity. Instead, to assess the filter performance, r is assumed to be a highly nonlinear, RFS-type function of the displacement u, i.e. of the spring elongation (Rose et al. 1981; Corigliano et al. 2006):

$$R(u) = a\,u\exp[-n\,u] \tag{2.47}$$

where a and n are unknown model parameters in need of tuning. Even if inspired by tight binding studies in atomistic simulations, law (2.47) is to be considered as phenomenological description of damaging processes taking place inside the spring: once a peak reaction is attained, softening (i.e. strength degradation) sets in and drives the state toward a smooth failure, occurring when $u \to +\infty$. Therefore, the two parameters a and n in (2.47) can be related to the strength r^M and the toughness G of the spring, through:

$$
\begin{aligned}
R^M &= \frac{a}{en} \\
G &= \int_0^\infty R\mathrm{d}u = \frac{a}{n^2}
\end{aligned}
\tag{2.48}
$$

where e is the Nepero number.

Law (2.47) can be handled as a tensile envelope, with damage activation/deactivation conditions to be adopted to properly describe unloading/reloading paths, see e.g. (Mariani and Ghisi 2007). In accordance with previous papers (Mariani 2009a, b), we instead assume here that damage evolution is captured by strength degradation only, and model (2.45) is managed as a holonomic (nonlinear elastic) law.

Fig. 2.1 Single degree-of-freedom structural system

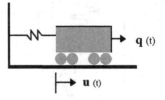

As aforementioned, we focus on pseudo-experimental (numerical) tests only. They consist in running direct analyses with known (target) values of model parameters, and then adding a white noise of assigned variance to the system output. This procedure allows us to obtain scattered measurements, which are then used to feed the filters.

In order to handle a stable system dynamics, followed by divergence (i.e. by $u \to +\infty$) due to the inception and growth of damage in the spring, the applied load $F(t)$ (see Eq. 2.47) has been assumed to monotonically increase in time according to:

$$F(t) = 0.5 + 0.0075t(\text{N}) \qquad (2.49)$$

see also (Corigliano and Mariani 2004). With the mass initially at rest, this loading condition allows the system to be stable up to $t \cong 150$ s; beyond this threshold, softening in the spring becomes dominant (i.e. the transmitted force gets vanishing), and displacement u diverges.

In the analyses, the mass has been assumed $m = 9.72 \text{ Ns}^2/\text{mm}$, see also (Corigliano and Mariani 2004). Measurements consist of the current mass displacement only, featuring a noise level characterized by a standard deviation $w = 0.05$ mm.

The results relevant to the tracking of the whole system state (i.e. u, $\dot{u}$ and $\ddot{u}$) are reported in Fig. 2.2, as obtained by running the EK-PF and, for comparison purposes, the PF and the S-PKF. In these plots, the dashed lines represent the target system response; the orange squared symbols are instead the discrete-time estimations furnished by the filters, and the blue circular symbols stand for the measurements (that are displacement values only). The figure illustrates that the three filters are all capable to track the initial, stable oscillations and the transition to the unstable regime due to inception of softening. Even if a high number of particles (500 in this analysis) has been adopted, the PF is not able to attain the same accuracy of the S-PKF; the EK-PF (run using 5 particles) is instead very accurate, performing slightly better than the S-PKF.

We now move to the system identification task. As usual [see, e.g. (Ljung 1999)], the following results have been obtained by setting the pivotal entries of $\boldsymbol{P}_0$ relevant to model parameters to be (at least) two orders of magnitude larger than those relevant to state variables. By this means, model calibration is enhanced, since information (actually, innovation) brought by measurements is trusted much more than the current estimates.

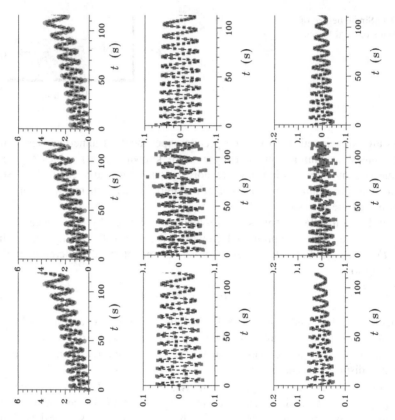

Fig. 2.2 State tracking. Comparison between target (*dashed lines*) and tracked (*orange squared symbols*) system evolution, in terms of: (*left column*) displacement u; (*central column*) velocity u̇; (*right column*) acceleration ü. Results obtained by running: (*top row*) EK-PF; (*middle row*) PF, and (*bottom row*) S-PKF

In terms of time evolution of the estimates of model parameters a and n, it is shown in Fig. 2.3 that they rapidly converge to the target values in the stable dynamic regime, independently of the initialization guess (here in the range between 50 and 150 % of the target values). The SPKF and the PF perform better than the EK-PF in the short-term time interval, featuring higher convergence rates without excessive oscillations of the estimates. However, as soon as the system stability threshold is approached, wild oscillations of increasing amplitude set in which leads to diverging model calibration furnished by SPKF and PF. On the contrary, the EK-PF does not show such wild oscillations, and always provides stable, unbiased estimates.

To obtain insights into the superior performance of the EK-PF, Figs. 2.4 and 2.5 report the projections onto the two model parameter axes of the time evolution of the (smoothed) distribution of particles deployed by PF and EK-PF, respectively. It can be seen that step #2 of prediction stage of the Table 2.6 proves to be very

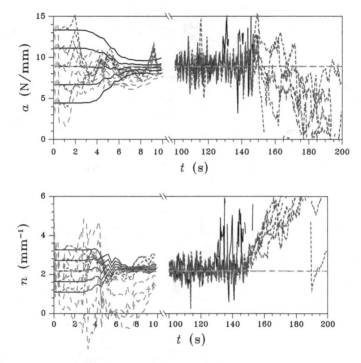

Fig. 2.3 Model calibration. Time evolution of estimated model parameters (*top row*) *a* and (*bottom row*) *n*, at varying initialization values. Results obtained by running: EK-PF (*long-dashed blue lines*), PF (*dashed orange lines*) and S-PKF (*continuous black lines*)

efficient in moving the particles toward the region of major interest, with distributions that are not spread over an extensive range of values. This eventually assists us to avoid divergence of the estimates.

Next, we study the performance of Bayesian filters for a slightly more difficult task: the dual estimation of a system having a bilinear constitutive model for its spring. The system is the same as before, but now the relationship between the force in spring R and the displacement u reads:

$$R = \begin{cases} k_1 u & if \quad u < u_M \\ k_1 u_M + k_2(u - u_M) & if \quad u > u_M \end{cases} \qquad (2.50)$$

where k_1 denotes initial slope of the constitutive model of the spring; u_M is the limit at which spring constitutive model starts its bilinear behavior; and k_2 denotes the gradient of force–displacement after the displacement has exceeded u_M.

The strength of the constitutive law (2.50) lies in the versatility in simulating three different material behaviors, namely the linear-hardening, linear-perfect plastic and linear-softening. Under monotonically increasing loadings, depending on the k_2 value this bilinear constitutive law can be adopted to deal with identification of parameters of a structure whose behavior may not be known as a priori.

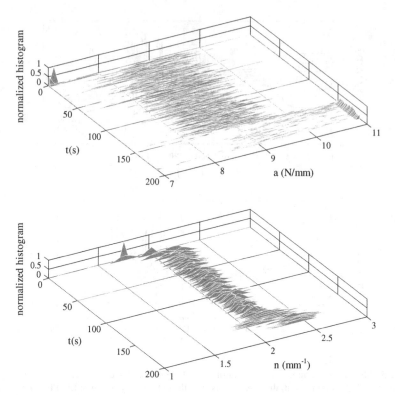

Fig. 2.4 PF, projections onto the parameters (*top*) *a* and (*bottom*) *n* axes of the evolution of particles

While dealing with joint state and parameter estimation, the main drawback of such constitutive law is the intricate interrelation of components of the state vector, when the parameter of the constitutive model are included into the state vector. Consider the state-space representation of the system, augmented state vector incorporates k_1, k_2 and u_M so as:

$$\vartheta = \begin{bmatrix} k_1 \\ k_2 \\ u_M \end{bmatrix}. \tag{2.51}$$

At each time iteration, the evolution equation, based on the value of u_M may find two different functional form: if displacement of the spring is less than u_M, only the initial linear behavior of the spring gets involved; if displacement of the spring exceeds u_M, nonlinearity of spring affects the spring force. Thus filter has to decide which path to follow as long as deterministic information is not available for u_M. In what follows, the results of application of nonlinear versions of Kalman filters and Particle filter and also a hybrid extended Kalman particle will be presented. The results are organized in three sets, each one of the filtering algorithms

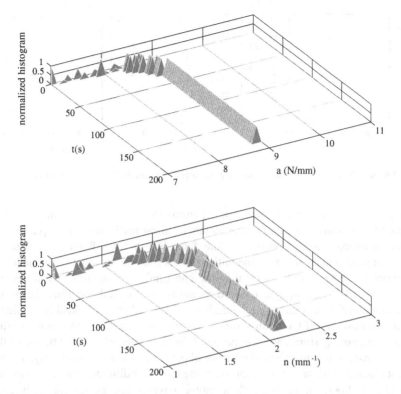

Fig. 2.5 EK-PF, projections onto the parameters (*top*) a and (*bottom*) n axes of the evolution of particles

is assessed when dealing with reference problems of each scenario: linear hardening, linear-perfectly plastic and linear-softening constitutive laws.

As aforementioned, in all the analyses, pseudo-experimental data are used instead of data coming from experiments; the numerical data contaminated by a zero mean additive white noise are therefore taken as observations of the system. The initial slope k_1 is always assumed to be 3.27 N/nm, while $k_2 = k_1/10$ for hardening, $k_2 = 0$ for plasticity and $k_2 = -k_1$ to mimic softening behavior. The value of the threshold of linear behavior u_M is set to 0.46 mm; the mass has been assumed $m = 9.72$ Ns^2/mm, see also (Corigliano and Mariani 2004; Eftekhar Azam et al. 2012a). Measurements consist of the current mass displacement only, featuring a noise level characterized by a standard deviation $w = 0.01$ mm. In order to incept a nonlinear behavior due to damage in the spring, the applied load q has been assumed to monotonically increase in time according to (2.49). Since the main objective of this study is the calibration of constitutive parameters, we just include the plots of parameter estimation unless there is a specific reason to present state estimate plots.

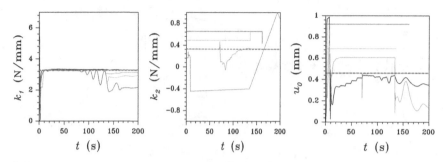

Fig. 2.6 Results of EKF for estimation of parameters of linear-hardening constitutive law

Figures 2.6, 2.7 and 2.8 show the performance of the EKF in simultaneous calibrating the three constitutive parameters of linear hardening, linear plastic and linear softening case, respectively. The filter is run for different initialization values; it is seen that except for the initializations from target values, in none of the scenarios the EKF is able to identify the constitutive parameters. As aforementioned, the EKF is a straight-forward extension of the Kalman filter, based on linearization of the evolution equation. It is suitably adopted for weakly nonlinear problems; however, if the nonlinearity is severe, such linearization is not accurate enough and poor performance is expected. It has to be underlined that tuning of the filter, in order to obtain unbiased estimate of parameters is not always easy, and we do not claim that we have tuned optimally the filters for different initializations and constitutive laws. In essence, three noise covariances associated with each parameter are tuning knobs of the system (Bittanti and Savaresi 2000). One has to notice that as the number of the parameters increases, their simultaneous tuning might become more difficult and algorithm appears to be practically inefficient.

Next, the results relevant to the performance of the SPKF are presented; even though SPKF has proved to outperform the EKF in many cases, it suffers from problem of positive definiteness of covariance matrix when dealing with parameter identification (Holmes et al. 2008), and also the tuning of the scale factor might become critical (Mariani 2009b). Figures 2.9, 2.10 and 2.11 present the results obtained by the SPKF when dealing with the three different scenarios of constitutive laws. Similar to the previous case, the filter is run with different initializations to see whether convergence is triggered from different starting points. It is seen that the performance of SPKF is quite poor, as it is not able to furnish unbiased estimates of the parameters, except for the case that the initial guess are set at the target values of parameters. We remind that in excess of three fictitious noise covariance to be tuned, within the SPKF algorithm also the scale factor should be tuned accurately; moreover, such a factor is used to allow the filter to capture local effects of nonlinearities of the evolution equation. Adding this to the three former parameters, one can see how delicate the task of tuning can become.

Since common extensions of the KF cannot furnish unbiased estimates of constitutive parameters, we make recourse to Particle filters as they are basically

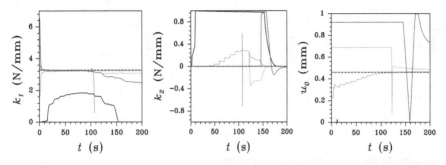

Fig. 2.7 Results of EKF for estimation of parameters of linear-plastic constitutive law

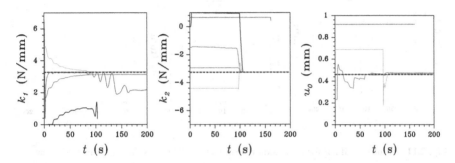

Fig. 2.8 Results of EKF for estimation of parameters of linear-softening constitutive law

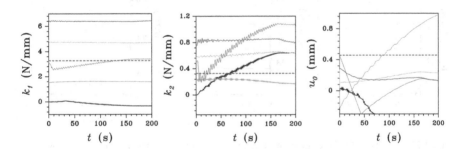

Fig. 2.9 Results of SPKF for estimation of parameters of linear-hardening constitutive law

designed for nonlinear systems with arbitrary uncertainty associated with them. Figures 2.12, 2.13 and 2.14 show the results of estimation of the parameters of linear-hardening, linear-perfect plastic and linear-softening constitutive model. Even though the particle filter is devised for nonlinear/non-Gaussian systems, it is seen through the graphs that it fails to estimate the parameters appropriately.

In designing a PF, it should be noticed that an appropriate initial guess of the distribution of the state of the system is essential to enhance the performance of the filter. Nevertheless, the value of the covariance of the noise for calibrating the

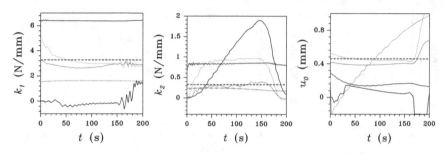

Fig. 2.10 Results of SPKF for estimation of parameters of linear-plastic constitutive law

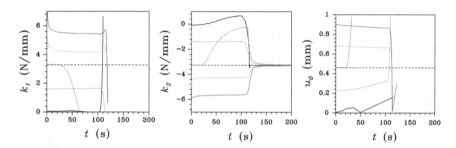

Fig. 2.11 Results of SPKF for estimation of parameters of linear-softening constitutive law

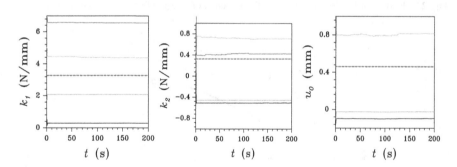

Fig. 2.12 Results of PF for estimation of parameters of linear-hardening constitutive law

parameters plays an important role (Arulampalam et al. 2002); moreover, they should be appropriately adjusted in order to let scattering of the samples in the feasible range of the parameter. We illustrate these issues via numerical examples. For ease of tuning, it is primarily assumed that we have quite reasonable a priori knowledge of k_1 and u_0 and aim to estimate only k_2. Figures 2.15, 2.16, 2.17, 2.18, 2.19 and 2.20 show the results of analysis for estimation of k_2. Looking at Figs. 2.15 and 2.18, they plot the time histories of estimation of the parameter k_2, supposing that the values of k_1 and u_M are a priori known. Moving from Figs. 2.15, 2.16, 2.17 and 2.18, we have changed the intensity if the tuning noise to highlight its

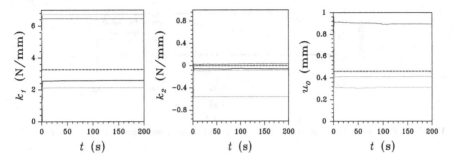

Fig. 2.13 Results of PF for estimation of parameters of linear-plastic constitutive law

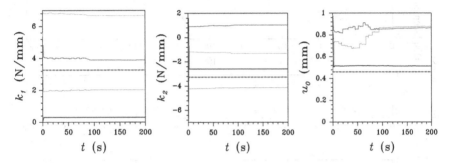

Fig. 2.14 Results of PF for estimation of parameters of linear-softening constitutive law

importance in the parameter estimation. In both cases the initial value of the parameter is set to 50 % of the target value. In the graph shown in Fig. 2.15, the value of the noise for tuning k_2 is set to 10^{-2} N^2/mm^2, which permit the evolution of the particles finally converge to the target value. On the contrary, the noise value equal to 10^{-4} N^2/mm^2 which is used to obtain the results shown in Figs. 2.18, 2.19 and 2.20, does not let the algorithm to sample efficiently, and the ensemble of the particles does not finally converge to the target values of the parameters.

To compare the performance of the particle filter when the tuning noise intensity varies, one can confront Figs. 2.16 and 2.19. At $t = 100$ s, as the parameter k_2 enters in the system evolution due to the inception of nonlinearity, for the case with the noise equal to 10^{-4}N^2/mm^2, estimates of the states of the system diverge, while in with the noise equal to 10^{-2}N^2/mm^2 states are estimated unbiasedly. This corroborates the idea that a small value for tuning noise intensity prevents the cloud of the particles to efficiently approximate the a posteriori distribution of the state. To investigate this issue in more details, we have focused on the histograms of the particles and their associated weights at $t = 130$ s, where there is a sharp change in the estimation of displacements (see Fig. 2.17). Looking at the histograms and particle weights shown in Fig. 2.20, it is seen that the cloud of the particles, shown via histogram, are far from the observation vicinity (the red

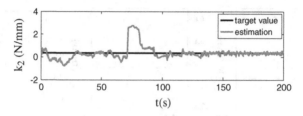

Fig. 2.15 Parameter estimates while noise covariance is set appropriately ($10^{-2}\,\mathrm{N}^2/\mathrm{mm}^2$)

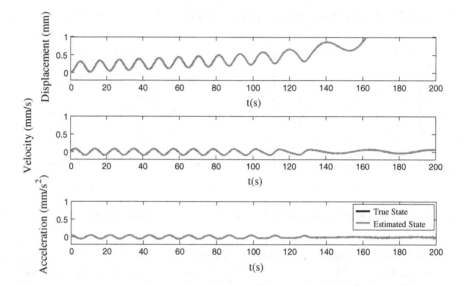

Fig. 2.16 State estimates when noise covariance is set appropriately ($10^{-2}\,\mathrm{N}^2/\mathrm{mm}^2$)

vertical line), where the distance of the closest bin to the observation is about 0.15 mm. As a consequence, in Fig. 2.20 all of the particles have found equal normalized weights; their distance from the observation vicinity is too far, as a consequence the associated probability with each particle becomes less than the round-off errors. On the contrary, looking at the same time instant in the case in which estimates are converging target values, it is seen that the distance of the closest been to the observation is about 0.004 mm; thus, in Fig. 2.18 the particles closer to observation have found a more significant normalized weight whereas other have smaller weights. Such diversity of weights shows that the particles are distributed in a zone which is close to the observation.

In what precedes, it has been shown that the proper choice of noise covariance has fundamental effects on the performance of PF. In case of dealing with one single parameter, it is not difficult to tune the filter; however, while dealing with more parameters, finding the right combination might become difficult. To address the

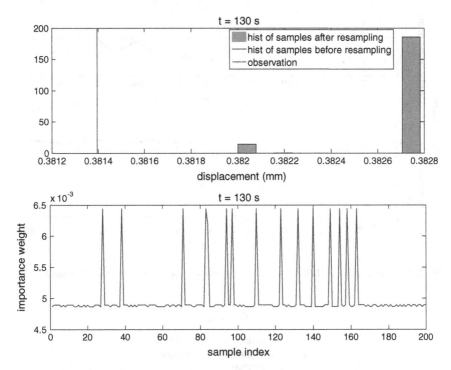

Fig. 2.17 Histogram of observable part of state vector (*top*) and associated sample weights (*bottom*) though through the *top figure* it seems that the sample has degenerated, through the bottom it is seen that many samples have significant weights. Also notice that samples are distributed in a close neighborhood of observation (*red vertical line*)

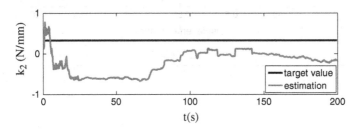

Fig. 2.18 Parameter estimates when noise covariance is not set appropriately ($10^{-4} \, \text{N}^2/\text{mm}^2$)

issues induced by simultaneous track of the three parameters shown in Figs. 2.11, 2.12, 2.13 and 2.14, for instance the step-function like behavior seen in Fig. 2.14 when calibrating u_0, we focus on the state estimation time histories, see Fig. 2.21, and consider the jump at $t = 34$ s. To have a closer look at what happens while this jump occurs, once again we make use of histogram of the distribution of the particles in two time instants: the beginning of the time step; the end of the time step. Before

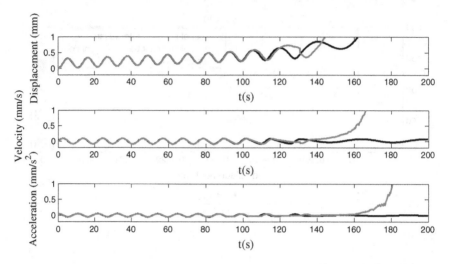

Fig. 2.19 State estimates when noise covariance is not set appropriately ($10^{-4}\,\mathrm{N}^2/\mathrm{mm}^2$)

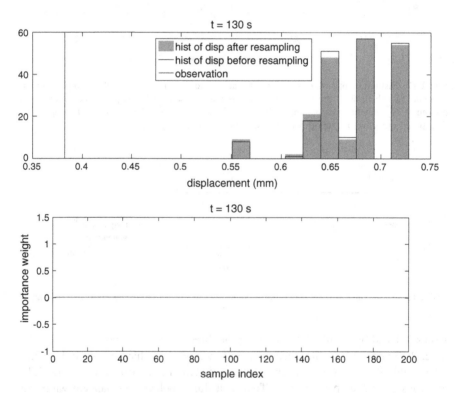

Fig. 2.20 Histogram of observable part of state vector (*top*) and associated samples weights (*bottom*) though from *top* it is seen that the sample cloud is quite far from observation neighborhood (*vertical red line*) consequently none of the particles find significant weights

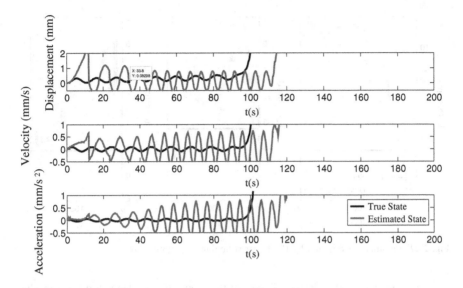

Fig. 2.21 State estimation by PF, linear softening constitutive law

proceeding with this objective, let us review again the particle filter algorithm. The procedure is triggered by drawing a number of N_P samples from a Gaussian distribution, then at each time instant t_k the same number of samples are drawn from transition prior. By transition prior, we mean a Gaussian distribution which its mean equals to the value of evolved estimated state at previous time step t_{k-1} while it's covariance equals to the covariance of the process noise. This procedure practically is equal to generation of N_P Gaussian random numbers, and adding to them the value of x_k which is evolved through evolution function. In the next stage, the probability of realization of each sample is computed. In this study, it is assumed that observation equation is contaminated by a white Gaussian process; hence, calculation of the probability of realization of each particle will be a function of a norm of the distance of the particle from the observation. The functional form of a multivariate Gaussian distribution reads as:

$$p(z) = \frac{1}{\sqrt{2\pi|\Sigma|}} e^{-\frac{1}{2}(z-\mu)^T \Sigma^{-1}(z-\mu)} \tag{2.52}$$

where μ and Σ denote mean and covariance of the state vector, respectively; $|.|$ stands for the determinant of the matrix. Within the PF algorithm, the above mentioned formula is used to compute the probability of realization associated with each particle $x_k^{(i)}$, according to :

$$p\left(y_k|x_k^{(i)}\right) = \frac{1}{\sqrt{2\pi|W|}} e^{-\frac{1}{2}\left(y_k - h_k\left(x_k^{(i)}\right)\right)^T W^{-1}\left(y_k - h_k\left(x_k^{(i)}\right)\right)}. \tag{2.53}$$

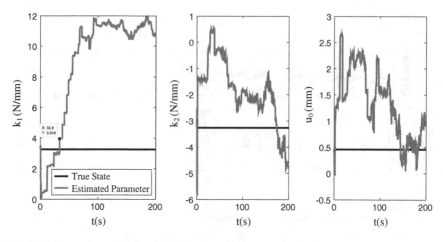

Fig. 2.22 Parameter estimation by PF, linear softening constitutive law

However, in case the observable part of the cloud of particles is too far from the observation y_k, the calculated probability will equal zero due to round off errors. To cope with ill-conditioning, it is set to a small value. As a result, all the particles will find an equal weight. In this condition, at the resampling stage, the resampled cloud will not change considerably, and will be similar to the previously existing cloud of particles. If the observable part of the cloud of particles approaches to observation vicinity (i.e. the zone in which at least a number of the probabilities are not affected by round-off error) a sharp change in the estimation of the state will occur. The gradient of such change in estimation of the observable part of state vector is obviously toward improvement in the estimate; however, the hidden (unobserved) part of state entries may or may not change in the direction to converge to an unbiased estimate, as seen in Fig. 2.22.

To visualize the phenomenon, the time evolution of displacement and parameters of the system are shown in the same plot, see Fig. 2.23. Now we regard a few time intervals of interest, and look at the histograms of particles at some time instants picked before and after the jump, we keep the time instant $t = 11.92$ s as reference instant.

In Fig. 2.24 it is seen that cloud of particles is not including the observation and the distance of the closest bin to the observation is about 0.2 mm (the value of the observation is indicated by a red vertical bar in the graph). Consequently, all the probabilities become zero, due to the round-off errors. To cope with the problem of ill-conditioning caused by the zero probabilities, in case of a zero probability, it is set to the smallest value that the computer program used accounts for it. That is, all the particles find the same weight. Figure 2.26 shows the histograms of k_1, k_2 and u_0 respectively. As a consequence of the equal weights of the particles; it is seen that, before and after resampling stage, the histograms are not changed (Fig. 2.25).

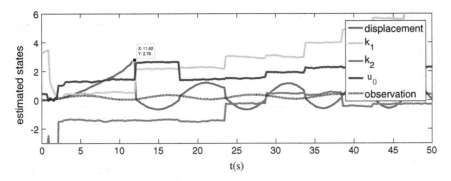

Fig. 2.23 State and parameter estimation by use of PF

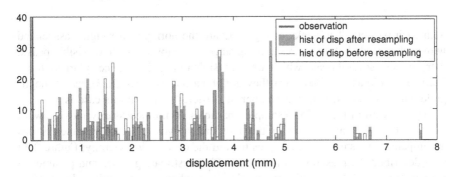

Fig. 2.24 Histogram of estimated displacements @ $t = 11.7$ s

Now let us look at $t = 11.92$, plots included in Figs. 2.27, 2.28 and 2.29 look considerably similar to previous time instant $t = 11.7$ s; however it seems that the cloud of samples is now closer to observation, as seen Fig. 2.27.

In what follows, histograms related to time instant t = 12.13 s are assessed. First see Fig. 2.30, in which the histogram of displacements is shown. Again, the red bar signifies the value observation y_k at related time instant, at its intersection with horizontal axis. It is seen that they are scattered throughout a wide interval; however, some particles have approached observation vicinity, as close as required to have non-zero weights for a couple of the particles, see Fig. 2.32.

To have a more clear idea, in Fig. 2.31 we have enlarged the vicinity of observation and histogram of resampled particles in order to highlight the changes in the particle cloud after resampling stage. We have to remark that the plot is an enlargement also in ordinate. It is clearly seen that a few particles (represented via blue histogram) have reached quite close to observation (red bar) so that their associated weight has become significant (see Fig. 2.32); as a consequence, in the resampling stage, the particles far from observation neighborhood are eliminated, and the ones close to it are duplicated. Figure 2.32 shows the weights associated

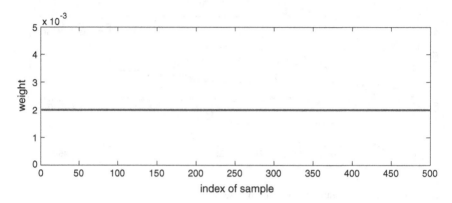

Fig. 2.25 Weights associated with each particle @ $t = 11.7$ s before resampling

with each particle. The peaks in Fig. 2.32 are the normalized weights associated with each particle before the resampling stage. The closer ones have visible peaks; there are also several peaks which are not visible in Fig. 2.32; once enlarged, those become visible as well; however they are about ten (see Fig. 2.33), nearly negligible when compared with the number of particles which in this case is 500.

As it is seen in Fig. 2.34, resampled particles do not necessarily move toward the target value; this is due to the fact that a wrong set of parameters has accompanied the shift of the samples toward the observation vicinity. Figure 2.34 well described the reason of failure of the PF in estimating states and parameters namely the distance of could of samples from observation vicinity. In order to alleviate such a problem, a remedy is to push the cloud of the samples toward observation vicinity. It can be done by employing the EKF: in each iteration, the EKF is used to update each particle by considering the information contained in the latest observation (de Freitas et al. 2000). More precisely, in the sampling stage, the samples are drawn from the transition prior; afterwards, each sample is updated by the EKF and so is pushed toward the observation vicinity. To some extent, this approach alleviates the problems arouse by choosing a suboptimal sampling distribution namely the transition prior. Figures 2.35, 2.36 and 2.37 show performance of a generic PF enhanced by the EKF. It is seen that such approach substantially improves the estimate of the parameters of the system.

To allow a clear understanding of the algorithm, let us look more closely at Fig. 2.37. Filter results from the initialization at 50 % of the target values is chosen just as an example. Figures 2.38 and 2.39 show the state and parameter estimation obtained through the EK-PF. It is seen that an excellent convergence is achieved. Figure 2.40 supports the idea that, by updating each individual particle within cloud of samples via EKF, the ensemble has to approach the zones of high probability.

As one can see in Fig. 2.40, after the EKF stage is implemented, the cloud of the samples drawn in the sampling stage moves toward the red bar (observation vicinity). In the resampling stage, the particles with higher probabilities are duplicated, and the ones with lower probability are eliminated; consequently, the

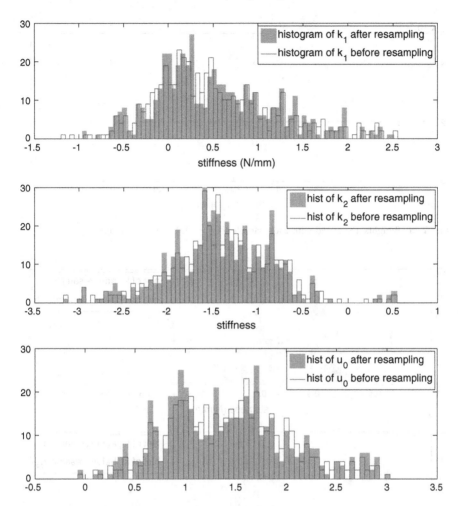

Fig. 2.26 Histogram of estimated parameters before and after resampling stage @ $t = 11.7$ s, *top* k_1; *middle* k_2; *bottom* u_0

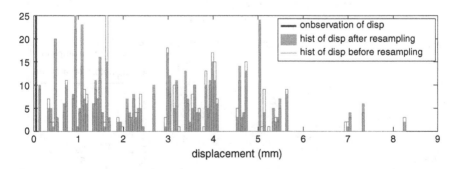

Fig. 2.27 Histogram of displacements @ $t = 11.92$ s

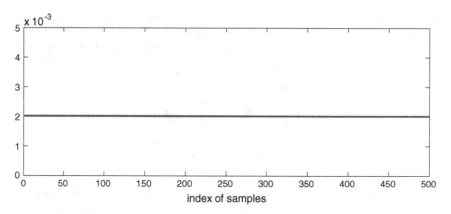

Fig. 2.28 Weights associated with each particle @ $t = 11.92$ s before resampling

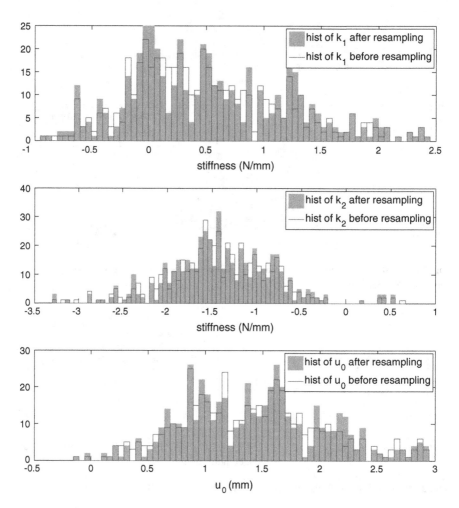

Fig. 2.29 Histograms of estimated parameters before and after resampling stage @ $t = 11.92$ s, *top* k_1; *middle* k_2; *bottom* u_0

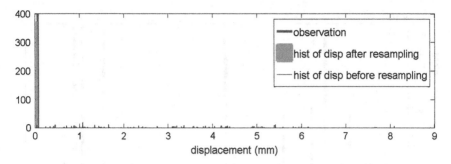

Fig. 2.30 Histogram of displacements @ $t = 12.13$ s

Fig. 2.31 Close up of histogram of displacements @ $t = 12.13$ s

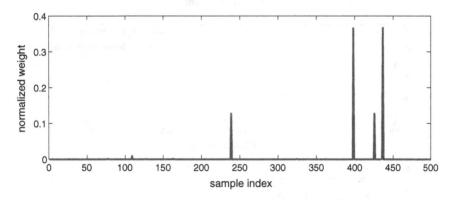

Fig. 2.32 Weights associated with each particle @ $t = 12.13$ s before resampling

cloud of the samples once again approaches the observation vicinity. Assessing other time instants always reveals the same results.

An extensive assessment of the performances of the Bayesian filters, when dealing with highly nonlinear dynamics of a SDOF system, has been presented. Though the studied mechanical system has only one degree-of-freedom, the

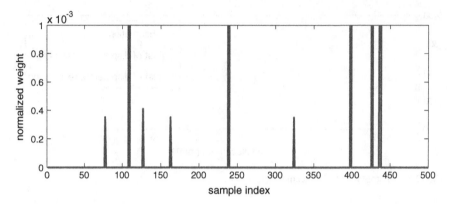

Fig. 2.33 Close up plot of weights associated with each particle @ $t = 12.13$ s before resampling

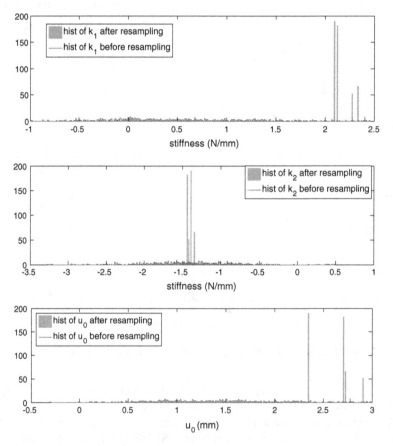

Fig. 2.34 Histogram of estimated parameters before and after resampling stage @ $t = 12.13$ s, *top k_1; middle k_2; bottom u_0*

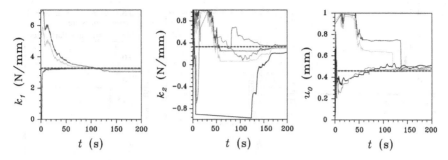

Fig. 2.35 Results of EK-PF for estimation of parameters of linear-hardening constitutive law

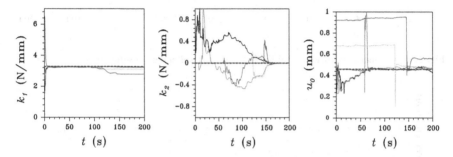

Fig. 2.36 Results of EK-PF for estimation of parameters of linear-plastic constitutive law

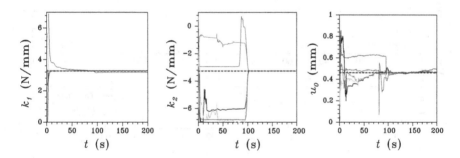

Fig. 2.37 Results of EK-PF for estimation of parameters of linear-softening constitutive law

extended state vector has three state components (displacement, velocity and acceleration) and 2 or 3 parameters (in case of a exponential softening constitutive law two parameters are to be calibrated, whereas in a bilinear one three parameters exist); consequently the extended state vector is multivariate even in present case. It was observed that the EKF, SPKF and PF all fail to furnish satisfactory results concerning identification of the parameters of the system, whereas EK-PF provides quite reasonable estimation of the states and parameters: for the exponential

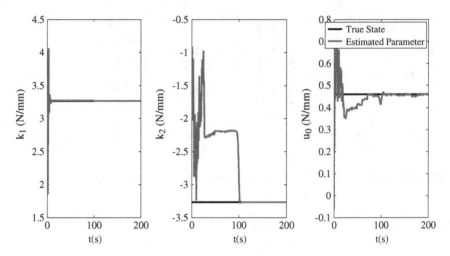

Fig. 2.38 Parameter estimation via EK-PF for a linear softening constitutive law

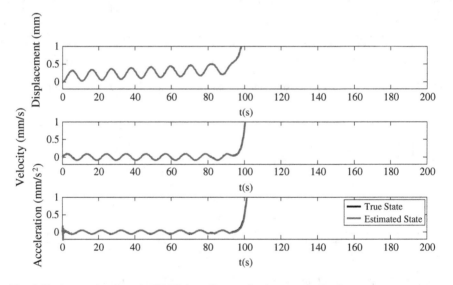

Fig. 2.39 State estimation via EK-PF for a linear softening constitutive law

behavior of the spring the results are unbiased for an extensive range of initializations; for the bilinear spring behavior EK-PF, in some cases, it converges to
unbiased solutions, and in some others, it converges to values affected by small
biases.

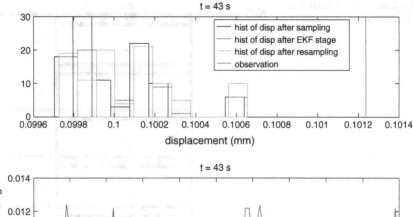

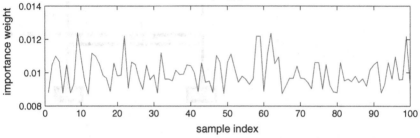

Fig. 2.40 *Top* histograms of displacement of the system at sampling (*black hist*), after EKF implemented on each sample (*magenta hist*) and after resampling stage (*green hist*), *bottom* associated importance weight with each particle

2.6.2 Multi Degrees-of-Freedom Dynamic System

In this Section, dual estimation of state and parameters of a shear type building is studied, as seen in Fig. 2.41. To start with the simplest case, we focus on the linear elastic response. By neglecting dissipating phenomena, the governing equations of motion thus is expressed as:

$$M\ddot{u} + Ku = F(t) \qquad (2.54)$$

where M and K denote the stationary mass matrix and stiffness matrix, respectively:

$$M = \begin{bmatrix} m_1 & & & \\ & m_2 & & \\ & & \ddots & \\ & & & m_n \end{bmatrix} \qquad (2.55)$$

Fig. 2.41 Schematic view of a shear building

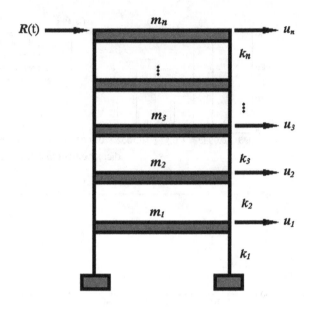

$$K = \begin{bmatrix} k_1 + k_2 & -k_2 \\ -k_2 & k_2 + k_3 \\ & & \ddots \\ & & & k_{n-1} + k_n & -k_n \\ & & & -k_n & k_n \end{bmatrix} \qquad (2.56)$$

whereas $F(t)$ is the external loading vector; in general, $F(t)$ can be any kind of loading. However in this study, we assume that it is a harmonic force applied to the top floor:

$$F(t) = \begin{bmatrix} 0 \\ \vdots \\ 0 \\ \varrho \sin \omega t \end{bmatrix} \qquad (2.57)$$

where ϱ and ω are the amplitude and the frequency of the excitation, respectively. To numerically solve (2.54), the Newmark explicit time integrator has been used, see Eqs. (2.40–2.44).

To write the equations in a discrete state-space form, we introduce an extended state z that, at each time instant t_k, includes u, $\dot{u}$ and $\ddot{u}$ according to:

$$z_k = \begin{bmatrix} u_k \\ \dot{u}_k \\ \ddot{u}_k \end{bmatrix}. \qquad (2.58)$$

The state-space form of (2.54) then reads:

$$z_k = A_k z_{k-1} + B_k \tag{2.59}$$

where

$$A_k = \begin{bmatrix} I - \beta \Delta t^2 M^{-1} K & \Delta t I - \beta \Delta t^3 M^{-1} K & -\beta \left(\frac{1}{2} - \beta\right) \Delta t^4 M^{-1} K + \Delta t^2 \left(\frac{1}{2} - \beta\right) I \\ -\gamma \Delta t M^{-1} K & I - \gamma \Delta t^2 M^{-1} K & -\gamma \left(\frac{1}{2} - \beta\right) \Delta t^3 M^{-1} K + \Delta t (1 - \gamma) I \\ -M^{-1} K & -\Delta t M^{-1} K & -\Delta t^2 \left(\frac{1}{2} - \beta\right) M^{-1} K \end{bmatrix} \tag{2.60}$$

and

$$B_k = \begin{bmatrix} \beta \Delta t^2 M^{-1} R_k \\ \gamma \Delta t M^{-1} R_k \\ M^{-1} R_k \end{bmatrix} \tag{2.61}$$

In this study, it is assumed that displacements and accelerations of the floors can be measured; thus the observation equation is written as:

$$y_k = H z_k + w_k \tag{2.62}$$

where H denotes a Boolean matrix of appropriate dimension, which links the observation process to the state of the system; w_k denotes the associated measurement noise; β and γ are parameters of the Newmark integration algorithm. For the dual estimation, the model parameter vector results:

$$\vartheta = \begin{bmatrix} k_1 \\ k_2 \\ \vdots \\ k_n \end{bmatrix}. \tag{2.63}$$

In the numerical analysis, we deal with a multiple-story shear building, featuring the same stiffness and mass values at each floor. We start by considering the smallest possible number of floors (say two), and see how many parameters are calibrated unbiasedly. In this regard, we assume $m_i = 25\,\text{kg}$ and $k_i = 300\,\text{kg/m}(i = 1 : n)$. The outcomes of state estimation and parameter calibration are a function of the quality and quantity of the information provided to the algorithms; by *quality,* we intend the accuracy of measurement devices, accuracy of the model of the system and initialization guess; by *quantity,* the number of degrees of freedom, whose evolution in time is measured, is intended.

This research focuses on the study of the effects of an increasing number of parameters in dual estimation of multi-dimensional mechanical systems. It has to be highlighted that the observable quantity is considered to be the displacement of the top floor only. Covariance of the measurement noise is assumed to be $2.7 \times 10^{-6}\,\text{m}^2$; the initial covariance of states (displacement, velocity and acceleration) is supposed to be very small ($10^{-10}\,\text{m}^2$), whereas diagonal entries of initial

Fig. 2.42 EKF (*red line*) and EK-PF (*blue line*) performances for calibration of a two-storey shear building stiffness's. The *black line* always represents the target value

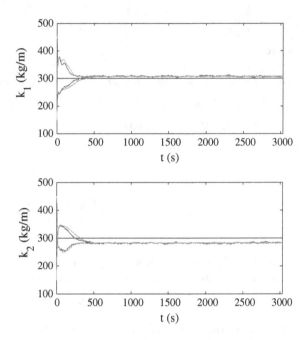

covariance of unknown parameters are assumed to be $10\,kg^2/m^2$. In all the analyses, the covariance of the fictitious noise for tuning the parameters is set to $10^{-3}\,kg^2/m^2$. Since states are always tracked unbiasedly, for the sake of brevity, the relevant results are not reported.

To ensure the algorithm has reached an unbiased estimate, it is a common practice to run analysis starting from different initializations; in case all converge to the same estimate, then it may be most likely an unbiased estimate. In this case, we initialize the analyses by values 50 % less and 50 % more than target value. We begin our numerical assessment by study of a two DOF structure and report the results of parameter estimation in Fig. 2.42: it is observed that two filters exhibit the same performances. In EK-PF procedure, 20 particles are deployed; by increasing the number of particles to 200, changes are visible in the plots of Fig. 2.42. Hence, the number of the particles was fixed to 20.

Though by increasing the number of particles toward infinity, particle filter can furnish unbiased estimates (Cadini et al. 2009), in practice, such a number of particles may be intractable for the current power of computational tools. By increasing the number of unknown parameters, it is observed that the bias in the estimates becomes more visible. In Fig. 2.43, it is seen that again both EKF and EK-PF exhibit the same performance; however, the bias in the estimates is increased when compared to a 2-DOF system. Moving to a 3-DOF and 4-DOF system, Figs. 2.43 and 2.44 reports the results when three and four inter-storey stiffnesses has to be estimated, respectively. Comparing with the case of a 2-storey shear building, again the bias in the estimate of the parameters increases.

Fig. 2.43 EKF (*red line*) and
EK-PF (*blue line*)
performances for calibration
of a three-storey shear
building stiffness's. The *black
line* always represents the
target value

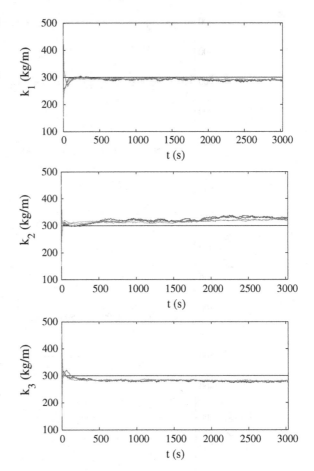

By exploring the literature concerning online methods for the identification of structures, one will see that most of it is focused on shear building structures with less than four stories [e.g. see (Chatzi et al. 2010; Gao and Lu 2006; Koh et al. 1995; Xie and Feng 2011)]. We avoid showing the results concerning estimation of more complicated structures, since they confirm the same trend observed in this reported part of the analysis. As the dimension of the state vector (hence the number of the parameters) increases, estimation of the parameters become increasingly difficult; in the jargon of dynamic programming, such a problem is termed *curse of dimensionality* (Bellman 1957). Powell (2007) illustrates this issue via an intuitive example: if state space has i dimensions and if each state component can take j possible values then we might have i^j possible states, i.e. by a linear increase in dimension of state vector, the dimension of the space of possibilities increases exponentially.

A possible remedy, for problems featuring high dimensionalities, is represented by searching for a possible subspace capturing the main variation in data; in

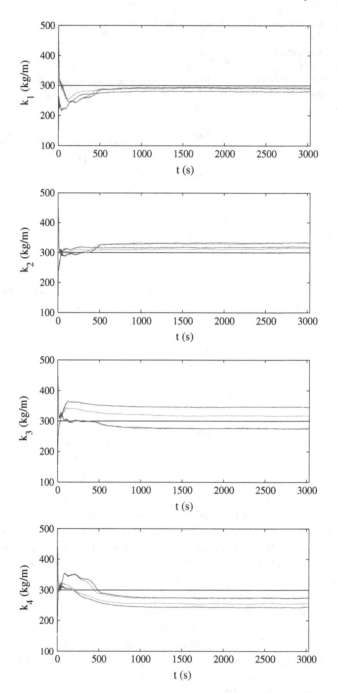

Fig. 2.44 EKF (*red line*) and EK-PF (*blue line*) performances for calibration of a four-storey shear building stiffness's. The *black line* always represents the target value

forthcoming Chapters, applicability of Proper Orthogonal Decomposition (POD) is primarily shown in constructing reduced order models, and afterwards such a model will be embedded in filtering schemes.

2.7 Summary and Conclusions

In this chapter, recursive Bayesian inference of partially observed dynamical systems has been reviewed. As a tool for structural system identification, nonlinear Bayesian filters are applied to dual estimation problem of linear and nonlinear dynamical systems. Dealing with a SDOF structure, it has been shown that the hybrid EK-PF filter is able to furnish a reasonable estimation of parameters of nonlinear constitutive models. Assessment of SDOF systems is followed by identification of multi storey buildings. In this regard, performances of the EK-PF and EKF algorithms are compared, and it has been concluded that they are nearly the same, and by an increase in the number of storeys of the building, the algorithms fail to provide an unbiased estimate of the parameters (stiffness of the storeys). Therefore, they are not reliable tools to monitor state and parameters of multi storey systems.

To develop a robust algorithm to monitor health of the structures via recursive Bayesian inference, we will make recourse to model order reduction of the dynamic systems. To this end, next Chapter reviews important features of proper orthogonal decomposition and its application to model order reduction of dynamic systems.

References

Adelino R, da Silva Ferreira (2009) Bayesian mixture models of variable dimension for image segmentation. Comput Methods Programs Biomed 94:1–14

Allen D, Darwiche A (2008) RC_Link: genetic linkage analysis using Bayesian networks. Int J Approximate Reasoning 48:499–525

Alvarado Mora MV, Romano CM, Gomes-Gouvêa MS, Gutierrez MF, Botelho L, Carrilho FJ, Pinho JRR (2011) Molecular characterization of the Hepatitis B virus genotypes in Colombia: a Bayesian inference on the genotype F. Infect, Genet Evol 11:103–108

Arulampalam MS, Maskell S, Gordon N, Clapp T (2002) A tutorial on particle filters for online nonlinear/non-Gaussian Bayesian tracking. IEEE Trans Sig Process 50:174–188

Bathe K (1996) Finite element procedures. Prentice-Hall Inc, Upper Saddle River

Bellman RE (1957) Dynamic programming. Princeton University Press, Princeton

Biedermann A, Taroni F (2012) Bayesian networks for evaluating forensic DNA profiling evidence: a review and guide to literature. Forensic Sci Int: Genet 6(2):147–157

Bittanti S, Savaresi SM (2000) On the parameterization and design of an extended Kalman filter frequency tracker. IEEE Trans Autom Control 45:1718–1724

Bittanti S, Maier G, Nappi A (1984) Inverse problems in structural elastoplasticity: a Kalman filter approach. In: Sawczukand A, Bianchi G (eds) Plasticity today. Applied Science Publications, London, pp 311–329

Cadini F, Zio E, Avram D (2009) Monte Carlo-based filtering for fatigue crack growth estimation. Probab Eng Mech 24:367–373

Caron F, Doucet A, Gottardo R (2012) On-line change point detection and parameter estimation with application to genomic data. Stat Comput 22:579–595

Chatzi EN, Smyth AW, Masri SF (2010) Experimental application of on-line parametric identification for nonlinear hysteretic systems with model uncertainty. Struct Saf 32:326–337

Corigliano A (1993) Formulation, identification and use of interface models in the numerical analysis of composite delamination. Int J Solids Struct 30:2779–2811

Corigliano A, Mariani S (2001a) Parameter identification of a time-dependent elastic-damage interface model for the simulation of debonding in composites. Compos Sci Technol 61:191–203

Corigliano A, Mariani S (2001b) Simulation of damage in composites by means of interface models: parameter identification. Compos Sci Technol 61:2299–2315

Corigliano A, Mariani S (2004) Parameter identification in explicit structural dynamics: performance of the extended Kalman filter. Comput Methods Appl Mech Eng 193:3807–3835

Corigliano A, Mariani S, Pandolfi A (2006) Numerical analysis of rate-dependent dynamic composite delamination. Compos Sci Technol 66:766–775

Creal D (2012) A survey of sequential Monte Carlo methods for economics and finance. Econ Rev 31(3):245–296

de Freitas JFG, Niranjan MA, Gee AH, Doucet A (2000) Sequential Monte Carlo methods to train neural network models. Neural Comput 12:955–993

Doucet A (1997) Monte Carlo methods for Bayesian estimation of hidden Markov models: application to radiation signals. (unpublished) doctoral dissertation, University Paris-Sud Orsay

Doucet A, Johansen AM (2009) A tutorial on particle filtering and smoothing: fifteen years later. Handbook of Nonlinear Filtering 12:656–704

Duan L, Gao W, Zeng W, Zhao D (2005) Adaptive relevance feedback based on Bayesian inference for image retrieval. Signal Process 85:395–399

Eftekhar Azam S, Mariani S (2012) Dual estimation of partially observed nonlinear structural systems: a particle filter approach. Mech Res Commun 46:54–61

Eftekhar Azam S, Ghisi A, Mariani S (2012a) Parallelized sigma-point Kalman filtering for structural dynamics, Comp Struct 92–93, pp. 193–205

Eftekhar Azam S, Bagherinia M, Mariani S (2012b) Stochastic system identification via particle and sigma-point Kalman filtering, Scientia Iranica A, 19:982–991

Gao F, Lu Y (2006) A Kalman-filter based time-domain analysis for structural damage diagnosis with noisy signals. J Sound Vib 297:916–930

Gelb A (1974) Applied optimal estimation. MIT Press, Cambridge

Gordon NJ, Salmond DJ, Smith AFM (1993) Novel approach to nonlinear/non-Gaussian Bayesian state estimation. IEE proceedings F, vol 140. pp 107–113

Hol JD, Schon TB, Gustafsson F (2006) On resampling algorithms for particle filtering. In: Proceedings of nonlinear statistical signal processing workshop 2006, pp 79–82

Holmes S, Klein G, Murray DW (2008) A square root unscented Kalman filter for visual monoSLAM. In: Proceedings—EEE international conference on robotics and automation, p 3710

Hughes TJR (2000) The finite element method. Linear static and dynamic finite element analysis. Dover, New York

Ishihara T, Omori Y (2012) Efficient Bayesian estimation of a multivariate stochastic volatility model with cross leverage and heavy-tailed errors. Comput Stat Data Anal 56(11):3674–3689

Ito K, Xiong K (2000) Gaussian filters for nonlinear filtering problems. IEEE Trans Autom Control 45:910–927

Jay E, Philippe Ovarlez J, Declercq D, Duvaut P (2003) BORD: Bayesian optimum radar detector. Signal Process 83:1151–1162

Julier SJ, Uhlmann JK (1997) New extension of the Kalman filter to nonlinear systems. Proceedings of SPIE—the international society for optical engineering, pp 182–193

Julier SJ, Uhlmann JK, Durrant-Whyte HF (1995) New approach for filtering nonlinear systems. In: Proceedings of the American control conference, pp 1628–1632

Julier S, Uhlmann J, Durrant-Whyte HF (2000) A new method for the nonlinear transformation of means and covariances in filters and estimators. IEEE Trans Autom Control 45:477–482

Kalman RE (1960) A new approach to linear filtering and prediction problems. J Basic Eng 82:35–45

Kitagawa G (1996) Monte Carlo filter and smoother for non-Gaussian nonlinear state space models. J Comput Graphical Stat 5:1–25

Koh CG, See LM, Balendra T (1995) Determination of storey stiffness of three-dimensional frame buildings. Eng Struct 17:179–186

Lazkano E, Sierra B, Astigarraga A, Martínez-Otzeta JM (2007) On the use of Bayesian Networks to develop behaviours for mobile robots. Rob Auton Syst 55:253–265

Li P, Goodall R, Kadirkamanathan V (2004) Estimation of parameters in a linear state space model using a Rao-Blackwellised particle filter. IEE proceedings: control theory and applications, vol 151. pp 727–738

Ljung L (1999) System identification. Theory for the user, 2nd edn. Prentice Hall, Englewood Cliffs

Mariani S (2009a) Failure of layered composites subject to impacts: constitutive modeling and parameter identification issues. In: Mendes G, Lago B (eds) Strength of materials. Nova Science Publishers, New York, pp 97–131

Mariani S (2009b) Failure assessment of layered composites subject to impact loadings: a finite element, sigma-point Kalman filter approach. Algorithms 2:808–827

Mariani S, Corigliano A (2005) Impact induced composite delamination: state and parameter identification via joint and dual extended Kalman filters. Comput Methods Appl Mech Eng 194:5242–5272

Mariani S, Ghisi A (2007) Unscented Kalman filtering for nonlinear structural dynamics. Nonlinear Dyn 49:131–150

Miazhynskaia T, Frühwirth-Schnatter S, Dorffner G (2006) Bayesian testing for non-linearity in volatility modeling. Comput Stat Data Anal 51:2029–2042

Mitra SK, Lee T, Goldbaum M (2005) A Bayesian network based sequential inference for diagnosis of diseases from retinal images. Pattern Recogn Lett 26:459–470

Powell WB (2007) Approximate dynamic programming: solving the curse of dimensionality. Princeton University Press, Princeton

Rose JH, Ferrante J, Smith JR (1981) Universal binding energy curves for metals and bimetallic interfaces. Phys Rev Lett 47:675–678

Saleh GMK, Niranjan M (2001) Speech enhancement using a Bayesian evidence approach. Comput Speech Lang 15:101–125

Ting J, D'Souza A, Schaal S (2011) Bayesian robot system identification with input and output noise. Neural Networks 63:99–108

Van der Merwe, R. 2004, Sigma-point Kalman filters for probabilistic inference in dynamic state-space models, Oregon Health and Science University

Velarde LGC, Migon HS, Alcoforado DA (2008) Hierarchical Bayesian models applied to air surveillance radars. Eur J Oper Res 184:1155–1162

White OL, Safaeinili A, Plaut JJ, Stofan ER, Clifford SM, Farrell WM, Heggy E, Picardi G (2009) MARSIS radar sounder observations in the vicinity of Ma'adim Vallis, Mars. Icarus 201:460–473

Xie Z, Feng J (2011) Real-time nonlinear structural system identification via iterated unscented Kalman filter. Mechanical Syst Signal Process 28:309–322

Yahya AA, Mahmod R, Ramli AR (2010) Dynamic Bayesian networks and variable length genetic algorithm for designing cue-based model for dialogue act recognition. Comput Speech Lang 24:190–218

Yang S, Lee J (2011) Predicting a distribution of implied volatilities for option pricing. Expert Syst Appl 38:1702–1708

Zhou H, Sakane S (2007) Mobile robot localization using active sensing based on Bayesian network inference. Rob Auton Syst 55:292–305

Chapter 3
Model Order Reduction of Dynamic Systems via Proper Orthogonal Decomposition

Abstract In this chapter, the performance of reduced order modeling of dynamic structural systems based on the proper orthogonal decomposition (POD) technique is investigated. Singular value decomposition and principal component analysis of the so-called snapshot matrix are considered to generate the reduced space, onto which the system equations of motion are projected to speed up the computations. To get intuitions into the achievable computational efficiency and the capability of POD to provide an input-independent reduced model, we consider the 39-story Pirelli tower in Milan-Italy. First, it is assumed that a shear model of the building is excited by the May 18-1940, Mw 7.1, El Centro earthquake, and the ensemble of the data necessary to build the reduced model is acquired. Then, the local and global accuracies of the same reduced model in tracking the dynamics of the building are assessed, if excited by the May 6-1976, Mw 6.4, Friuli earthquake and by the January 17-1995, Mw 6.8, Kobe earthquake, which differ from the El Centro one in terms of excited vibration frequencies. It is shown that POD allows to attain a speedup close to 250, when the reduced order model is required to retain a high fidelity.

3.1 Introduction

While working with a space discretized system, proper orthogonal decomposition (POD) automatically seeks a dependence structure between the degrees-of-freedom, which are normally assumed to be independent (Buljak 2012). This is accomplished through a set of ordered, orthonormal bases, and through information regarding the relevant energy contents. The POD has been developed independently by various researchers in different fields (see e.g. Kosambi 1943; Karhunen 1947; Obukhov 1954) and has been called by utilizing variety of names. When applied to finite dimensional systems, it is called principal component analysis (PCA) (Jolliffe 1986) which originates in the Pearson's (1901) study on plane and line fitting to point sets. When working with distributed parameter systems, it is called Karhunen–Loève decomposition (KLD); nevertheless, its discrete representation is introduced as well (Fukunaga 1990). Another POD

S. Eftekhar Azam, *Online Damage Detection in Structural Systems*,
PoliMI SpringerBriefs, DOI: 10.1007/978-3-319-02559-9_3,
© The Author(s) 2014

technique is called singular value decomposition (SVD) (Mees et al. 1978), innovation of such technique is credited to Eckart and Young; where they proposed extension of eigen value decomposition for general non square matrices (Klema and Laub 1980). For a detailed proof of equivalency of PCA, KLD and SVD readers are referred to (Liang et al. 2002a).

As a result of standard numerical tools developed to extract proper orthogonal modes (POMs) of the systems, and as a result of its power in feature extraction and reduced modeling, presently the POD is extensively employed in various engineering fields. To illustrate this issue further, one can see that the POD has been applied for reduced order modeling of heat transfer phenomena (Samadiani and Joshi 2010) and other field such as: computational fluid dynamics (Smith et al. 2005; Tadmor et al. 2006), micro electro mechanical systems (Liang et al. 2002b), various fields of computational physics (Lucia et al. 2004) and aeroelasticity (Thomas et al. 2003). The method of POD has obtain great reputation in the field of structural dynamics, where it is employed for active sensing (Park et al. 2008) and active control of structures (Al-Dmour and Mohammad 2002), damage detection (De Boe and Golinval 2003; Galvanetto et al. 2008; Shane and Jha 2011c), model updating (Lenaerts et al. 2003; Hemez and Doebling 2001), modal analysis (Han and Feeny 2003; Feeny 2002) and model reduction (Steindl and Troger 2001; Buljak and Maier 2011). For a more comprehensive examination of the related literature, readers are referred to (Kerschen et al. 2005). The study carried out in the literature recommends that the POD is a strong tool for model order reduction of structural systems; however, the research lacks a specific study of expediting, computational accuracy of the reduced model and robustness to the change in the source of excitation. The study presented in this chapter acknowledges those aforementioned issues. The author of this monograph has coauthored a journal article on the topics covered in this chapter (Eftekhar Azam and Mariani 2013).

Next, in the Sect. 3.2, structural dynamics of systems are examined and studied in this chapter as well; moreover, their associated set of governing differential equation and the numerical scheme are employed for time discretization. Section 3.3 reviews fundamentals of the POD. Afterwards, the fundamental studies carried out in finding the links between POMs and eigen modes of linear structures are summarized in Sect. 3.4. Furthermore, a reduced model is constructed via Galerkin projection of the set of governing equations onto the reduced space spanned by POMs which is presented in Sect. 3.5. Finally, the results of the numerical assessment of efficiency of POD: speedup and accuracy of reduced models of Pirelli tower, as a case study, are investigated and reported in Sect. 3.6.

3.2 Structural Dynamics and Time Integration

In this study, the POD for reduced order modeling of dynamic systems is exploited. Subsequently, such reduced model will be embedded into a Bayesian filter in the forth-coming chapters. In this section, the differential equations of the governing dynamics of structural systems are reviewed herein; moreover, the numerical integration scheme employed for time discretization of the aforementioned differential equations is briefly discussed.

The dynamic response of the structural system to the external loads is allowed to be described by the following linear equations of motion:

$$M\ddot{u}(t) + D\dot{u}(t) + Ku(t) = F(t) \tag{3.1}$$

where: M is the mass matrix; D is the viscous damping matrix; K is the stiffness matrix; F is the time-dependent external force vector; $\ddot{u}$, $\dot{u}$ and u are the time-varying vectors of accelerations, velocities and displacements, respectively. For instance, in a shear model of a building (such as the one adopted in Sect. 2.6), these vectors gather the lateral displacements, velocities and accelerations of the storeys.

Equation (3.1) is usually arrived once the structural system has been space discretized (e.g. through finite elements), or once assumptions concerning the behavior of the building (e.g. shear-type deformation) have taken into account. This preliminary stage of the analysis can affect the sparsity of matrices in Eq. (3.1), and can therefore have an impact on the speedup obtained through the POD as well.

In this study, the solution of the vectorial differential equation (3.1) is advanced in time by utilizing the Newmark explicit integration scheme. For details, the reader is referred to Sect. 2.6.

3.3 Fundamentals of Proper Orthogonal Decomposition for Dynamic Structural Systems

The aim of reduced order modeling is to automatically find a solution for the following two conflicting requirements: create the smallest possible numerical model of the original dynamic system; preserve accuracy in the description of the system behavior. Standard techniques attempt to extract fundamental features from the dynamic model; thus the governing equations can be thereafter projected onto a reduced state space or subspace.

The POD, in its snapshot version (Sirovich 1987), is adopted to build the model-specific optimal linear subspace on the basis of an ensemble of system observations in this study. Let us consider the displacement vector $u \in \mathbb{R}^m$, $\mathbb{R}$ being the set of real numbers and m the dimension of vector u; we assume that u effectively describes system evolution (i.e. it does not need to be supplemented by $\dot{u}$ and $\ddot{u}$ to define the full state space), and consider a set of arbitrary orthonormal bases $\{\varphi_i\}$, $i = 1, \ldots, m$, spanning its vector space $\mathbb{R}^m$. Such bases satisfy $\varphi_i^T \varphi_j = \delta_{ij}$ ($j = 1, \ldots, m$), where δ_{ij} is the Kronecker's delta (such that $\delta_{ij} = 1$ if $i = j$, otherwise $\delta_{ij} = 0$). The original vector u can then be written as a linear combination of the aforementioned bases, according to:

$$u = \sum_{i=1}^{m} \varphi_i y_i = \Phi y \qquad (3.2)$$

where y_i are the combination coefficients, arranged in the column vector y, and:

$$\Phi = [\varphi_1 \varphi_2 \ldots \varphi_m] \qquad (3.3)$$

is the matrix gathering all the bases.

To ensure computational gain, we define a reduced representation of the state via:

$$u_l = \sum_{i=1}^{l} \varphi_i y_i = \Phi_l \alpha \qquad (3.4)$$

where we enforce $l < m$ or, for large systems, even $l \ll m$. In (3.4), Φ_l is the matrix gathering the first l columns of matrix Φ (i.e. the first l bases), and α collects the relevant first l components of vector y. The goal of POD is to provide an ordered sequence of the bases φ_i, so as to satisfy the following extreme value problem:

$$\min \|u - u_l\| \qquad (3.5)$$

where $\|\blacksquare\|$ represents the L^2 norm of vector. Given l, Eq. (3.5), thus it is required to find the optimal subspace spanned by the bases $\varphi_1, \ldots, \varphi_l$.

We now need to establish l on the basis of the required accuracy of the solution provided by the reduced order model, and to compute the bases gathered by Φ_l. Both problems can be attacked through the so-called snapshot version of POD. First, since we have to provide a subspace for the state vector u, the characteristic displacements $u^{(k)} = u(t_k)$ $(k = 1, \ldots, n)$ at n time instants are computed and collected in an ensemble, or snapshot matrix U, according to:

$$U = \left[u^{(1)} u^{(2)} \ldots u^{(n)} \right]. \qquad (3.6)$$

Next PCA and SVD, two POD methods for extracting so-called POMs are briefly discussed.

3.3.1 Principal Component Analysis

To detect the main dependence structure in an ensemble of data, PCA looks for the subspace which is able to keep the maximum variability in the data. A very naïve justification of this procedures is expressed as: in the state-space, the directions along which data vary are important, since the dynamics of the system is actually occurring along those directions, whereas the directions featuring no variations are redundant in the dynamic representation, and computational cost will be spent in calculating something that we already know if they were retained in the analysis.

Consider the aforementioned vector $u \in \mathbb{R}^m$; suppose $y_1, y_2, \ldots, y_m \in \mathbb{R}$ are the first, second,... and mth principal components, respectively. Let the first principal component y_1 be a linear combination of each element of the original vector, i.e.:

$$y_1 = \sum_{i=1}^{m} \xi_{i1} u_i = \xi_1^T u \tag{3.7}$$

where: $\xi_1 = \{\xi_{11}, \xi_{21}, \ldots, \xi_{m1}\}^T$. The variance of y_1, assumed to be a random variable, is then:

$$S_{y_1}^2 = \xi_1^T Z_u \xi_1 \tag{3.8}$$

where Z_u is the covariance of the variable u, assumed to be random as well. To find the direction in which the maximum variability of data is captured, we look for the direction in which the projection of the samples onto it yields the maximum variance. The maximum of $S_{y_1}^2$ will not be achieved for a finite value of ξ_1, thus a constraint have to be imposed and reads:

$$\max_{\xi_1} (\xi_1^T Z_u \xi_1), \ s.t. \ (\xi_1^T \xi_1) = 1. \tag{3.9}$$

Introducing the Lagrangian multiplier λ_1, from (3.8) and (3.9) we obtain:

$$L(\xi_1, \lambda_1) = \xi_1^T Z_u \xi_1 + \lambda_1 (1 - \xi_1^T \xi_1) \tag{3.10}$$

where $L(\blacksquare)$ is Lagrangian operator. After differentiation (3.10) gives:

$$\frac{\partial L(\xi_1, \lambda_1)}{\partial \xi_1} = 2(Z_u - \lambda_1 I)\xi_1 = 0 \Rightarrow Z_u \xi_1 = \lambda_1 \xi_1 \tag{3.11}$$

where λ_1 and ξ_1 are the eigenvalue and the corresponding eigenvector of the covariance matrix Z_u, respectively.

Applying the same procedures, the objective function to be maximized in order to extract the principal components of a random variable which is written as:

$$\max_{\xi_i} (\sum_{i=1}^{m} \xi_i^T Z_u \xi_i), \ s.t. \ (\xi_i^T \xi_j) = \delta_{ij} \tag{3.12}$$

and the approximation error due to a representation by its first l principal components, $u \approx \sum_{i=1}^{l} y_i \xi_i$, would be:

$$\varepsilon^2(l) = E(\|u - u(l)\|^2) = \sum_{i=l+1}^{m} E(y_i^2) = \sum_{i=l+1}^{m} S_{y_i}^2. \tag{3.13}$$

One has to handle the covariance matrix of the random vectorial variable in order to compute the principal components. However, since in practical problems, it is usually impossible to determine this covariance matrix, it is a common

Fig. 3.1 Building the matrix
of snapshots

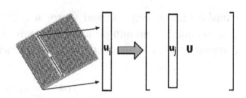

practice to use the correlation matrix as an acceptable approximation of it
(Schilders 2008). To approximate the covariance matrix with the required accuracy, one needs an appropriately chosen ensemble of the samples; such a seed of
samples is the so-called snapshot matrix, wherein each snapshot represents the
state of the system at a specific time instant (see Fig. 3.1).

The covariance of the data set, allocated in a snapshot matrix U, is then calculated as (Schilders 2008):

$$Z_u = \lim_{n \to \infty} \left(\tilde{Z}_u = \frac{1}{n} U U^T \right). \tag{3.14}$$

3.3.2 Singular Value Decomposition

Exploiting the singular value decomposition of the snapshot matrix U we obtain
(Liang et al. 2002a):

$$U = L \Sigma R^T \tag{3.15}$$

where: L is a $m \times m$ orthonormal matrix, whose columns are the left singular
vectors of U; Σ is a $m \times n$ pseudo-diagonal and semi-positive definite matrix,
whose pivotal entries Σ_{ii} are the singular values of U; R is a $n \times n$ orthonormal
matrix, whose columns are the right singular vectors of U.

The whole basis set Φ, i.e. the set of all the so-called POMs, is given by L, i.e.
by the left singular vectors of the snapshot matrix (Kerschen and Golinval 2002).
If singular values Σ_{ii} are sorted decreasingly, and the columns of L and R are
accordingly arranged; the decomposition (3.15) is such that the first l columns
(with l given) of $\Phi = L$ represent the optimal basis subset that fulfills (3.5).
Moreover, it is known (see, e.g. Kerschen and Golinval 2002) that the ith singular
value squared (i.e. Σ_{ii}^2) represents the maximum of the relevant oriented energy[1];
this means that the ith oriented energy is maximized, among all the possible unit

---

[1] The oriented energy of a vector along a direction is given by the magnitude of the projection of
the (m-dimensional) vector itself onto the mentioned direction, namely by the dot product of the
two vectors. When we have to deal with a vector sequence like U, the oriented energy of
the sequence is given by the sum of the magnitudes of the projections of all the vectors $u^{(k)}$ onto
the same direction.

vectors, by the basis φ_i. Since we are looking for the most informative subspace, which should be able to furnish as much insight as possible into the dynamics of the original system and therefore, into how energy fluxes take place inside, we retain in the reduced order model the proper modes φ_i that feature the highest singular values. Additional proper modes, featuring less energy contents, will be redundant in the reduced order representation, and add computational costs with marginal enhancement in the accuracy.

Now, having established a method to sort bases φ_i, and the link between the singular value Σ_{ii} and the energy content of the proper mode φ_i, we need to set l. According to (Kerschen and Golinval 2002), we assign the required accuracy p of the reduced order solution, intended as a fraction of the total oriented energy of the full model, and select the dimension l of the subspace by fulfilling:

$$\frac{\sum_{i=1}^{l} \Sigma_{ii}^2}{\sum_{i=1}^{m} \Sigma_{ii}^2} \geq p \tag{3.16}$$

hence, on the basis of the ratio between the sum of the singular values of the kept modes and the sum of all the singular values.

3.4   Physical Interpretation of Proper Orthogonal Modes

It is known that the POD is a statistical technique which extracts POMs from the response of the system. However, a close relationship has been established between the POMs and natural eigen-modes of a mechanical system (Feeny and Kappagantu 1998; Kerschen and Golinval 2002). The effort toward establishing a link between the POMs and eigen-modes of the system intends in making the POD a modal identification tool (Yadalam and Feeny 2011). To accomplish this task, theoretical and experimental study has been carried out to link the POMs with eigen-modes of a linear (Feeny 2002) and nonlinear (Georgiou 2005) mechanical systems. In this section, we do not discuss the details offered by the existing literature and only mean to summarize interesting findings published therein.

Free vibrations of an undamped linear system with mass matrix proportional with identity matrix (e.g. a shear building with equal masses at each storey) results in a set of POMs that asymptotically converge to eigen-modes of the system. POMs of a lightly damped similar system are reasonable approximations of eigen-modes of the system (Kerschen and Golinval 2002); however in case of forced harmonic vibration, there is no guarantee that POMs converge to eigen-modes.

When the system resonates at a certain frequency, independently of mass matrix entries, the POMs coincides with the respective eigen-modes of that frequency (Kerschen et al. 2005). It has been shown that POMs coincide with eigen-modes for many noise driven oscillators (Preisendorfer 1979); moreover, North has established a general criterion for symmetry of POMs and eigen-modes of the mechanical systems excited by noise (North 1984).

3.5 Galerkin Projection

Once POD has furnished the required subspace, the displacement vector can be approximated through $\mathbf{u}_l$. Since matrix $\boldsymbol{\Phi}_l$ is a function of the position vector only, and defines the shapes of POMs for the structure, while $\boldsymbol{\alpha}$ governs the evolution in time of the structural response, it follows that:

$$\ddot{u} \approx \boldsymbol{\Phi}_l \ddot{\boldsymbol{\alpha}} \quad \dot{u} \approx \boldsymbol{\Phi}_l \dot{\boldsymbol{\alpha}} \quad u \approx \boldsymbol{\Phi}_l \boldsymbol{\alpha}. \tag{3.17}$$

The equations of motion (3.1), allowing for (3.17), can now be approximately stated as:

$$M\boldsymbol{\Phi}_l \ddot{\boldsymbol{\alpha}}(t) + D\boldsymbol{\Phi}_l \dot{\boldsymbol{\alpha}}(t) + K\boldsymbol{\Phi}_l \boldsymbol{\alpha}(t) \cong F(t). \tag{3.18}$$

By defining the residual r of such approximation as:

$$r = F(t) - (M\boldsymbol{\Phi}_l \ddot{\boldsymbol{\alpha}}(t) + D\boldsymbol{\Phi}_l \dot{\boldsymbol{\alpha}}(t) + K\boldsymbol{\Phi}_l \boldsymbol{\alpha}(t)) \tag{3.19}$$

within a Galerkin projection frame (Steindl and Troger 2001), we enforce it to be orthogonal to the subspace $\boldsymbol{\Phi}_l$ spanned by the solution, i.e.:

$$\boldsymbol{\Phi}_l^{\mathrm{T}} r = 0. \tag{3.20}$$

Hence, the equations of motion of the reduced order model turn out to be:

$$\boldsymbol{\Phi}_l^{\mathrm{T}} M \boldsymbol{\Phi}_l \ddot{\boldsymbol{\alpha}}(t) + \boldsymbol{\Phi}_l^{\mathrm{T}} D \boldsymbol{\Phi}_l \dot{\boldsymbol{\alpha}}(t) + \boldsymbol{\Phi}_l^{\mathrm{T}} K \boldsymbol{\Phi}_l \boldsymbol{\alpha}(t) = \boldsymbol{\Phi}_l^{\mathrm{T}} F(t) \tag{3.21}$$

or, equivalently:

$$M_l \ddot{\boldsymbol{\alpha}}(t) + D_l \dot{\boldsymbol{\alpha}}(t) + K_l \boldsymbol{\alpha}(t) = F_l(t). \tag{3.22}$$

Once the solution of (3.22) is obtained, the full state of the system can be computed by utilizing (3.17).

3.6 Results: Reduced-Order Modeling of a Tall Building Excited by Earthquakes

For linear systems, it will be beneficial if POMs φ_i depend only on physical and geometrical properties of the structure, with marginal effects of the kind of loading considered in the phase of construction of the snapshot matrix. Since different loading conditions may excite a different set of structural vibration modes, what claimed here above does not necessarily hold true. Although a thorough analysis of theoretical aspects of the POD, when applied to structural systems, has been carried out in the literature, only a handful of studies are available on several practical points including the load-dependency of the POMs. Such issue may become crucial especially when the structure is subject to seismic loadings, which are difficult to predict in nature.

Fig. 3.2 The Pirelli tower in Milan, Italy

The performance of POD has been already assessed in defining reduced models for multi-support structures subject to seismic excitation (Tubino et al. 2003); furthermore, the POD has been applied for efficient reduced modeling of high-rise buildings subject to earthquake loads (Gutiérrez and Zaldivar 2000; Aschheim et al. 2002). However, its efficiency for high fidelity reduced order modeling of multi-storey buildings trained by a certain seismic load and excited by another one has not been done yet. In this section, we investigate whether a reduced order model, built by considering a specific input while constituting the snapshot matrix, can be used to represent with a similar level of accuracy the dynamics of the full structure in case of different excitation, in terms of e.g.: frequency content and therefore, excited vibration modes.

In the forthcoming numerical examples, we will set $p > 0.99$ to ensure accuracy. As a case study, we investigate the capability of POD in speeding up the computations by considering the Pirelli Tower in Milan, see Fig. 3.2. The building features 39 stories, and its total height is about 130 m. The plan dimensions of the standard floor are approximately 70×20 m. The structure is entirely made of CIP reinforced concrete. The structure is assumed to behave elastically, with lumped masses at each storey that basically undergo horizontal displacements. Such an assumption may be far from reality if the rigid diaphragm assumption does not hold true for vertical displacements of all the nodes at the same floor.

We start with a three-dimensional finite element discretization of the whole building featuring 6219 DOFs (Barbella et al. 2011). For the sake of simplicity, we have neglected the damping effect; hence, in a relative frame moving with the basement of the tower, the undamped equations of motion of the structure are written as:

$$\boldsymbol{M\ddot{u}} + \boldsymbol{Ku} = -a(t)\boldsymbol{MB} \tag{3.23}$$

where $a(t)$ denotes the earthquake-induced acceleration time history, whereas $\boldsymbol{B}$ is a Boolean matrix of appropriate dimension which defines the shacked DOFs. To simplify the problem, static condensation has been adopted to keep out the vertical displacements of the floors. By partitioning the nodal displacements $\boldsymbol{u}$ into horizontal $\boldsymbol{u}_h$ and vertical $\boldsymbol{u}_v$ components, we can write:

$$\begin{bmatrix} \boldsymbol{M}_h & 0 \\ 0 & \boldsymbol{M}_v \end{bmatrix} \begin{bmatrix} \boldsymbol{\ddot{u}}_h \\ \boldsymbol{\ddot{u}}_v \end{bmatrix} + \begin{bmatrix} \boldsymbol{K}_{hh} & \boldsymbol{K}_{hv} \\ \boldsymbol{K}_{vh} & \boldsymbol{K}_{vv} \end{bmatrix} \begin{bmatrix} \boldsymbol{u}_h \\ \boldsymbol{u}_v \end{bmatrix} = -a(t) \begin{bmatrix} \boldsymbol{M}_h\boldsymbol{B}_h \\ \boldsymbol{M}_v\boldsymbol{B}_v \end{bmatrix}. \tag{3.24}$$

Keeping only the horizontal DOFs only in the equations of motion, to be thereafter managed by POD, we arrive at:

$$\boldsymbol{M}_h\boldsymbol{\ddot{u}}_h + \left[\boldsymbol{K}_{hh} - \boldsymbol{K}_{hv}\boldsymbol{K}_{vv}^{-1}\boldsymbol{K}_{vh}\right]\boldsymbol{u}_h = -a(t)\boldsymbol{M}_h\boldsymbol{B}_h \tag{3.25}$$

where now $\boldsymbol{u}_h \in \mathrm{R}^{39}$.

To obtain the reduced model, the building has been assumed to be shaken by the well-known El Centro earthquake, whose time versus acceleration record, together with its relevant fast Fourier transform, is reported in Fig. 3.3. To give an idea about the number of vibration modes that may be excited by such earthquake, the first natural eigen-frequencies of the structure (see also Table 3.1) are denoted by red vertical lines in Fig. 3.3b. It can be deduced that only the first five eigen-modes of structure can be effectively excited as the power of the spectra of the accelerogram is intuitively seen to be small for the frequencies higher than the 6th natural frequency of the structure.

A comparison among the dynamics of the original 39-DOF system and the responses of reduced order models at varying accuracy index p (see Eq. (3.20)) has been performed. The link between p and the retained DOFs in the reduced systems is reported in Table 3.2. The result reported in Figs. 3.4 and 3.5 compare the time histories of (lateral) displacements, velocities and accelerations of the 20th and 39th (roof) floors, respectively, with the target values which are available from the simulations. In these plots, the blue vertical line indicates the end of the time window within which the snapshots are collected; hence, only around $t = 4$ s all the reduced order analyses start departing from the full model response.

To have a more clear view of the time histories, a close up of the last 5 s of the time histories of 20th floor is presented in Fig. 3.6. By making a comparison between time histories of displacements, velocities and accelerations, it can be observed that two POMs are sufficient for a reduced model to accurately reproduce displacements of the full model; however, at least four POMs are necessary to feature the same level of accuracy for velocities and accelerations as well. By investigating the FFTs of the aforementioned time histories (see Figs. 3.7, 3.8 and 3.9), it is shown that in the FFT of the displacement time histories, only the two

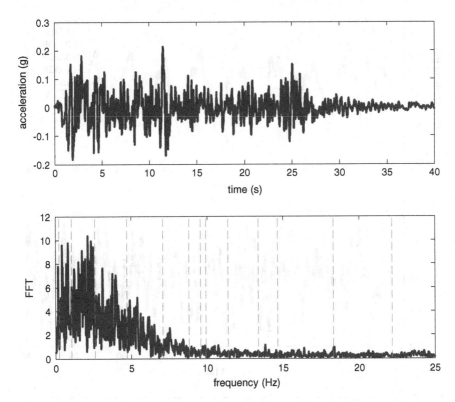

Fig. 3.3 *Top* May 18-1940, El Centro accelerogram (east–west direction); *bottom* relevant FFT

Table 3.1 First natural frequencies of the building

Vibration mode index	1	2	3	4	5	6	7	8	9	10	11	12	13
Natural frequency (Hz)	0.26	1.09	2.61	4.71	7.07	8.79	9.56	9.92	11.38	13.36	14.64	18.30	22.14

Table 3.2 Outcomes achieved through POD, in terms of accuracy p and speedup as functions of the number of DOFs retained in the reduced order model

# DOFs	p	Speedup
1	0.99	515
2	0.999	385
3	0.9999	276
4	0.99999	244

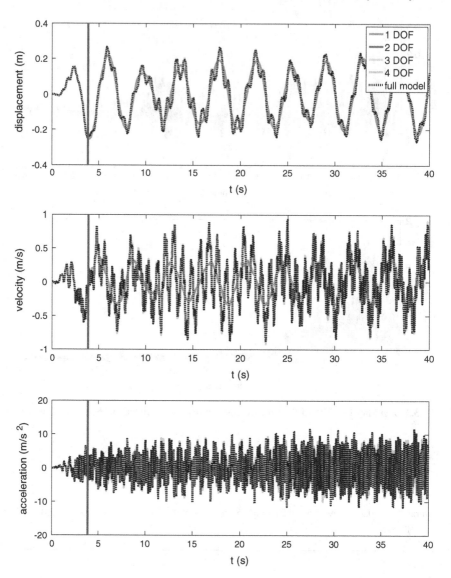

Fig. 3.4 Time histories of the horizontal, displacement (*top*), velocity (*middle*) and acceleration (*bottom*) of the 20th floor, as induced by the El Centro earthquake

first natural modes are effectively excited. Instead, in the velocity and acceleration time histories, looking at the FFTs, it is observed that the six and seven first natural frequencies are effectively excited. Such a trend suggests that a reduced model that retains a few POMs may feature a better accuracy in reconstruction of the displacements of the system, when compared with velocities and acceleration responses.

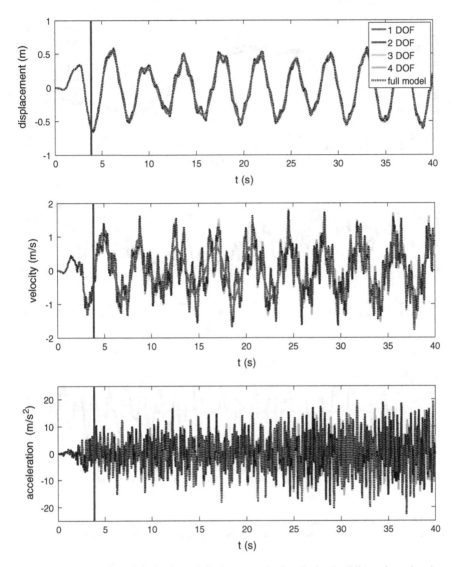

Fig. 3.5 Time histories of the horizontal displacement (*top*), velocity (*middle*) and acceleration (*bottom*) of the 39th floor, as induced by the El Centro earthquake

Moving to the speedup obtained by reducing the order of the full model, the results here discussed have been obtained with a personal computer featuring and Intel Core 2 Duo CPU E8400, with 4 Gb of RAM, running Windows 7 × 64 as operating system and performing the simulations with MATLAB version 7.6.0.324. The speedup values reported in Table 2.1 testify the dramatic decrease of the computing time obtained through POD, and reveal how powerful this methodology can be to approach real-time computing.

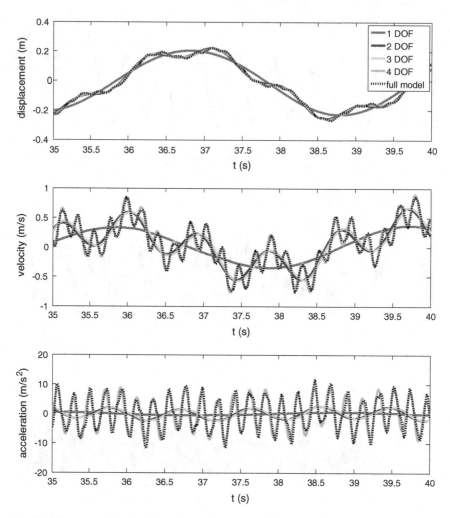

Fig. 3.6 Close up of the time histories of the horizontal displacement (*top*), velocity (*middle*) and acceleration (*bottom*) of the 20th floor, as induced by the El Centro earthquake

Previous figures have reported the results concerning time histories of two representative storeys of the structure: 20th storey is the mid floor and 39th storey is the last floor (roof) of the building. To further test the efficiency of the reduced models in reconstructing snapshots of the system, and therefore assess the capacity of the methodology in tracking the dynamics of all the system DOFs, two time instants are selected to assess the accuracy: Fig. 3.10a and b show snapshots taken in $t = 10$ s and $t = 30$ s of the analysis. At $t = 30$ s, the deformation of the structure is rather similar to a line with constant slope that is the reduced model with two POMs can reconstruct the relevant snapshot; however more POMs are required to appropriately approximate snapshot taken at $t = 10$ s, since the shape

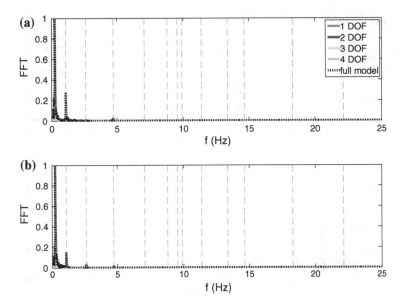

Fig. 3.7 FFTs of the horizontal displacements of the storeys as induced by the El Centro earthquake at **a** 20th and **b** 39th floors

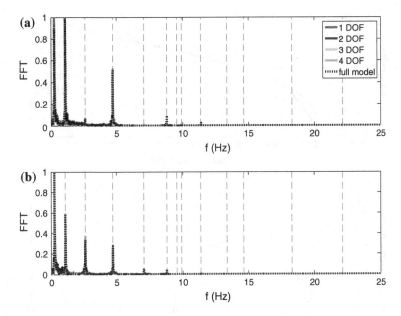

Fig. 3.8 FFTs of the horizontal velocities of the storeys as induced by the El Centro earthquake at **a** 20th and **b** 39th floors

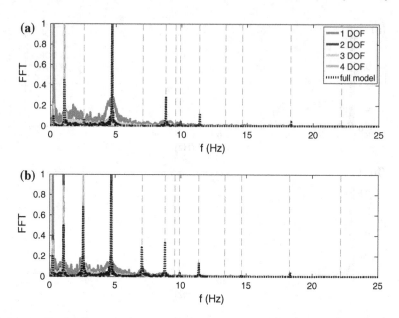

Fig. 3.9 FFTs of the horizontal accelerations of the storeys as induced by the El Centro earthquake, **a** 20th (*top*), and **b** 39th floor

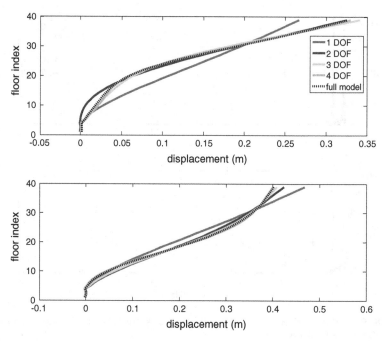

Fig. 3.10 Snapshots of the horizontal storey displacements as induced by the El Centro earthquake. *Top* t = 10 s and *bottom* t = 30 s

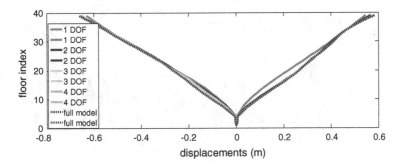

Fig. 3.11 Envelope of horizontal storey displacements, as induced by the El Centro earthquake

of the building is more complicated and higher modes are playing more significant role when compared to $t = 30$ s.

Another global feature of the reduced model which may be of interest for design practice is the envelope of the displacement, which is reported in Fig. 3.11. It is observed that even the reduced model with a single POM has an acceptable performance in reconstructing the envelope, even though it underestimates the envelope itself. By increasing the flexibility of the reduced model through additional POMs, as the higher POMs are retained in the analysis, it is observed that the envelope of the reduced model nearly matches that of full one.

To assess the efficiency of the reduced models in retaining the energy of the system, we now compare the resulting time histories of kinetic and potential energies of the system (see Fig. 3.12), respectively defined as: $E_k = \frac{1}{2}\dot{u}^T M \dot{u}$ and $E_p = \frac{1}{2}u^T M u$ for the full model; $E_{kl} = \frac{1}{2}\dot{\alpha}^T M_l \dot{\alpha}$ and $E_{pl} = \frac{1}{2}\alpha^T K_l \alpha$ for ROMs. The cumulative discrepancy of the energies of the reduced models from the target one is considered as well (see Fig. 3.13). It is seen that the energy time histories of the 4-DOF reduced model well match those of the full model. To have more insight into the ability of the models to preserve energy of the system, the cumulative discrepancies of kinetic and potential energies are reported as well. It is seen that as the number of the DOFs of the reduced model increases the slope of the relevant line decreases, it means the rate of accumulation of the discrepancy decreases. Besides, it is observed that the accumulation of the discrepancies features a line with an almost constant slope implies that at different time intervals of the analysis, the amount of energy loss is the same. It means that the rate of energy loss is constant; hence, the accuracy of the reduced model in terms of energy preservation is constant over the interval of the analysis.

From this point on, we examine the accuracy of the reduced models that are built via snapshots resulting from excitation by the El Centro earthquake, when the building are shaken by other seismic records. In this regard, as an instance, we consider the May-1976 Friuli earthquake which time history of its acceleration records along with the relevant FFT are shown in Fig. 3.14. To have an idea concerning the number of natural frequencies that are covered by this seismic action, again the red vertical lines (as indicator of the natural frequencies of the

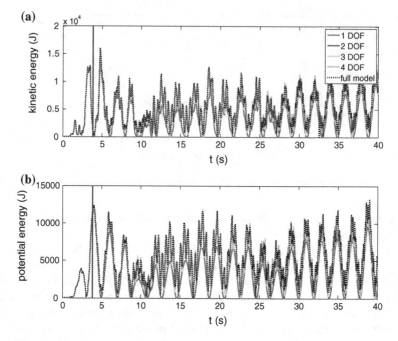

Fig. 3.12 Time histories of **a** kinetic and **b** potential energies

structure) are drawn in the figure to allow for understanding the number of eigen-modes which is excited by accelerogram of the relevant earthquakes. By an intuitive comparison of Figs. 3.3 and 3.14, it is observed that a different amount of eigen-modes of the structure are excited by the two earthquake records.

Let us now consider the time histories of displacement, velocity and acceleration of the 39th storey (see Fig. 3.15). It is seen that, while a 2-DOF reduced model satisfactorily mimics the behavior of the full model in terms of displacement, a 4-DOF reduced model is required to match the full model in terms of velocity and acceleration time histories. The number of POMs required for reconstructing the whole state of the structure, when it is shaken by Friuli earthquake, is the same as if it was shaken by El Centro earthquake. This fact shows that a reduced model built by the POD may be robust to change in the excitation source.

By investigating the FFTs of the aforementioned reported time histories (see Fig. 3.16a), the trend seen in the time histories of the state reconstruction is corroborated: one can see there are several peaks in the displacement response of the structure, when shaken by Friuli record; similarly to the FFTs of the structure when subjected to El Centro record, moving from displacement to velocity and acceleration FFTs, the number of peaks increases. Therefore, the number of POMs is required to match the FFT of the response of the structure increases.

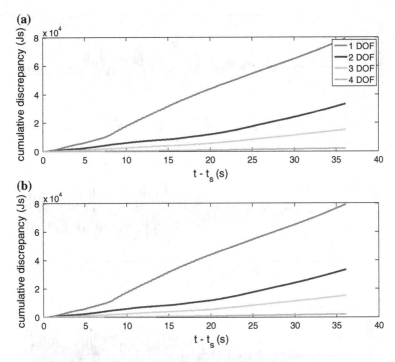

Fig. 3.13 Time evolution of cumulative discrepancy between full model and reduced order model, in terms of **a** kinetic, **b** potential energies

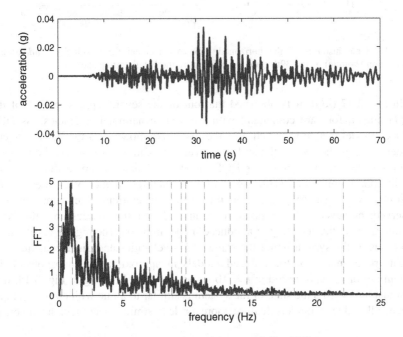

Fig. 3.14 *Top* May 6-1976, Friuli earthquake and *bottom* relevant FFT

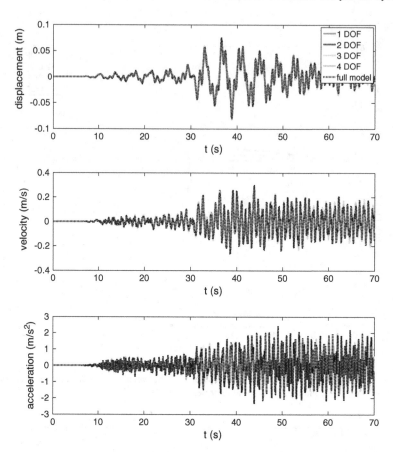

Fig. 3.15 Time histories of the horizontal floor displacement (*top*), velocity (*middle*) and acceleration (*bottom*) of the 39th floor, as induced by the Friuli earthquake

In Fig. 3.17 (top), it is observed that out of the several spikes in FFT of the displacement, four are coincident with the system natural frequencies. As DOF reduced model is able to match only the first spike, however a two DOF reduced model matches the two of the spikes relevant to natural frequencies, the reduced models featuring three and four DOF models are matched up to third and the fourth spikes coinciding with the third forth natural frequency of the system, respectively. Considering the velocities and accelerations, the same trend is observed; however in latter cases, more natural vibration modes are effectively excited. Hence, the accuracy of a reduced model in reconstructing the acceleration responses of the system is not the same as the velocities and displacements.

The performance of the reduced models in approximating snapshots of the system are once again tested at $t = 10$ s and $t = 30$ s. Looking at Fig. 3.18, it is seen that at $t = 10$ s the state of the system is similar to a line with constant slope; hence, all reduced models feature more or less similar accuracy; however, at

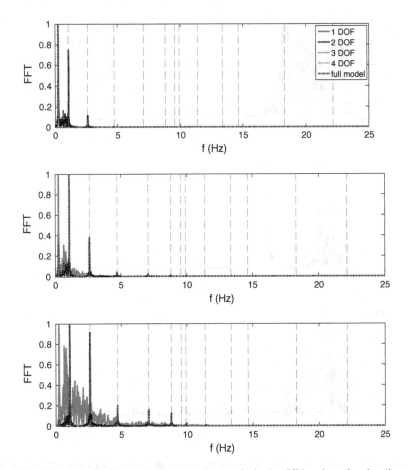

Fig. 3.16 FFTs of the 39th floor, displacement (*top*), velocity (*middle*) and acceleration (*bottom*) as induced by the Friuli earthquake

$t = 30$ s the state of the structure is more complicated and at least four POMs are required to approximate the considered snapshot.

Concerning the envelope of the displacements (see Fig. 3.19), it is observed that even a two DOF reduced model is matched with the envelope featured by the full model. It is observed that, in the vicinity of the 25th floor, there is a break in the envelope of the structure, while in the envelope of floor displacements relevant to the El Centro earthquake such a break is not observed. This is due to the fact that the range of frequency content of Friuli earthquake is wider than that of El Centro earthquake, see Figs. 3.3b and 3.14b, it results in excitation, and therefore contribution of higher natural modes in the response of the structure and as a consequence the shape of the structure may become more complicated.

To evaluate the accuracy of the reduced models concerning the energies, accumulated discrepancies has been considered; as previously, the time histories

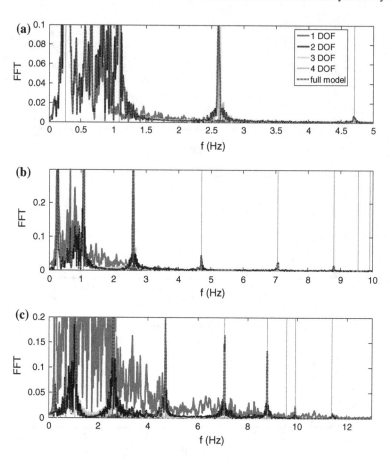

Fig. 3.17 Close up of FFT of the horizontal displacement (**a**), velocity (**b**) and acceleration (**c**) of 39th floors, as induced by the Friuli earthquake

feature the same features of those related to El Centro record. Figure 3.20 shows the accumulated discrepancies of kinetic and potential energies for two scenarios: the continuous lines represent the case in which snapshots are related to the El Centro excitation, instead the dot lines stand for the case in which snapshots are related to Friuli record. It is worth recalling that in both cases, the reduced and full models are shaken by Friuli record. It is seen that, despite the fact that the reduced models are constructed by different inputs in simulations, the accumulated discrepancies nearly coincide. However in this case, the accumulated discrepancies appears to be bilinear: the graphs look similar to an straight line which changes its slope at $t = 30$ s. This is due to the fact that the amplitude of the excitations increases at the vicinity of the $t = 30$ s, the increase in the energy of input excitation therefore changes the rate of accumulation of the discrepancies changes.

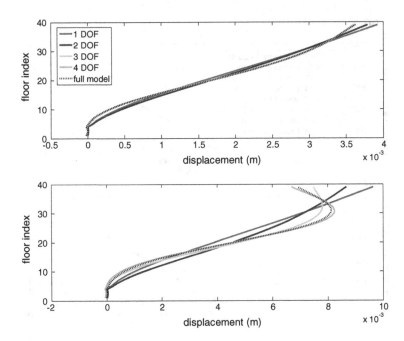

Fig. 3.18 Snapshots of the horizontal storey displacements at (*top*) t = 10 s, and (*bottom*) t = 30 s, as induced by the Friuli earthquake

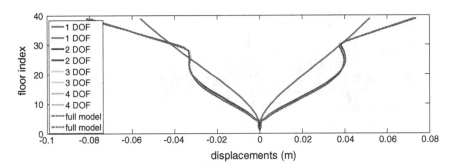

Fig. 3.19 Envelope of horizontal storey displacements, as induced by the Friuli earthquake

To conclude this section, we further assess the performance of the already built reduced models when excited by January 17-1995 Kobe acceleration record. In Fig. 3.21, the acceleration time history and its relevant FFT is presented. Once more, one can observe that the frequency content of this record is different from those of El Centro and Friuli. Figure 3.22 presents the time histories of displacement, velocity and acceleration of 39th storey. The situation is rather similar to the two previous cases: concerning displacements, reduced models retaining two or more DOFs nearly coincide with the full model, whereas dealing with velocity

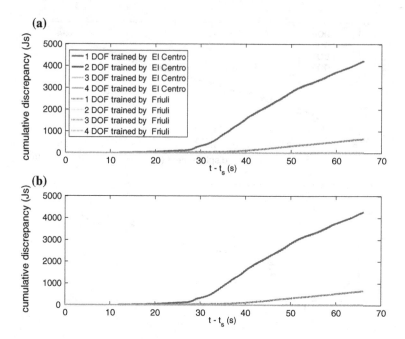

Fig. 3.20 Time evolution of cumulative discrepancy between full model and reduced order model, in terms of (**a**) kinetic, and (**b**) potential energies. Comparison between outcomes of the reduced order models trained by El Centro and Friuli earthquakes

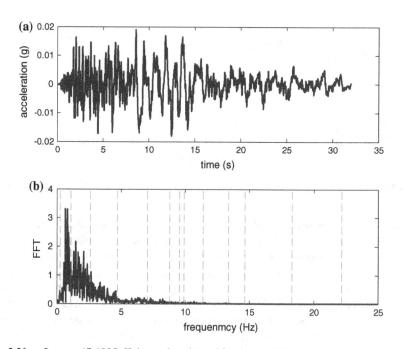

Fig. 3.21 **a** January 17-1995, Kobe earthquake and **b** relevant FFT

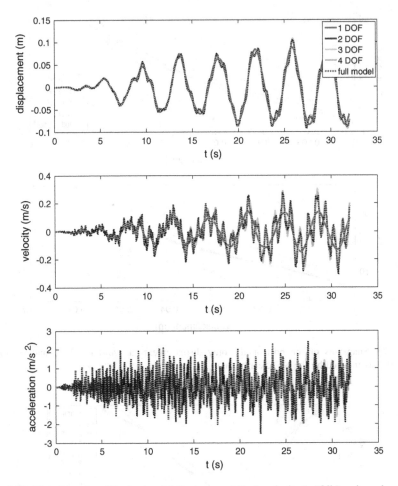

Fig. 3.22 Time histories of the horizontal displacement (*top*), velocity (*middle*) and acceleration (*bottom*) of the 39th floor, as induced by the Kobe earthquake

and acceleration a four DOF model is necessary to fully capture the dynamics of the system (Fig. 3.23).

Considering snapshots and envelope of the displacements of the system (see Figs. 3.24 and 3.25), a reduced model consisting of a single DOF is not able to feature the dynamics of the system similar to the case shaken by El Centro and Friuli earthquake. To assess the global efficiency of the reduced model when subject to Kobe record (see Fig. 3.25), once more, one can observe that the ability to retain energy is independent of the training stage. The reduced models have the same number of DOFs, no matter how many snapshots are collected from simulation of El Centro or Kobe earthquake simulations, which nearly feature the same level of accuracy.

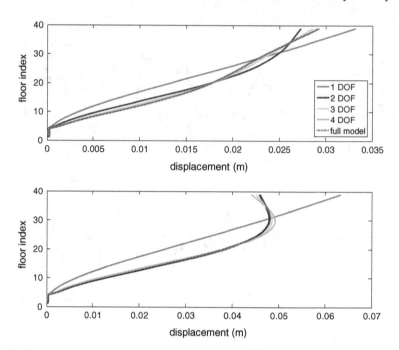

Fig. 3.23 Snapshots of the horizontal storey displacements at (*top*) t = 10 s, and (*bottom*) t = 30 s, as induced by the Kobe earthquake

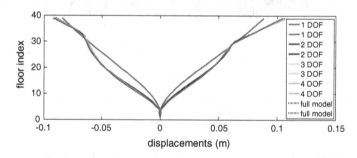

Fig. 3.24 Envelope of horizontal storey displacements, as induced by the Kobe earthquake

Through the results shown in this Section, it has been revealed that prediction capabilities of POD-based reduced order models when dealing with different seismic excitations along with their high speed-up in computation makes them suitable candidates for models used in online and real-time structural health monitoring.

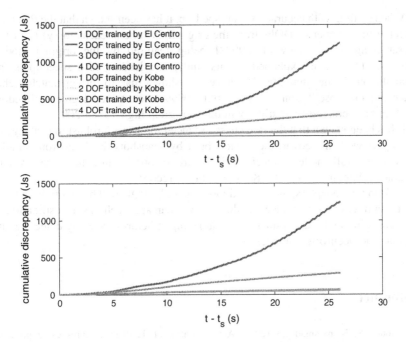

Fig. 3.25 Time evolution of cumulative discrepancy between full model and reduced order model, in terms of (*top*) kinetic, and (*bottom*) potential energies. Comparison between outcomes of the reduced order models trained by El Centro and Kobe earthquakes

3.7 Summary and Conclusion

In this chapter, we have investigated the capability and efficiency of the POD in reducing the order of dynamic structural systems. In its SVD description, the POD is expected to find the directions in which retain the maximum energy of the system, whereas its PCA explanation is based on the search for the directions which preserve the maximum variability of the set of samples, which are collected into the so-called matrix of snapshots. Handling snapshots collected in an initial time window, we have built the reduced model through a coupling of POD and Galerkin projection.

To assess the performance of the studied methodology, the Pirelli Tower in Milan has been assumed to be shaken by an earthquake. Concerning accuracy issues, time histories of the state of the system (storey displacements, velocities and accelerations), together with their associated Fourier transform, have been compared with their real values available through the simulations. The power of the order reduction method in preserving the energies of the system is tested via a comparison of their time histories with those of full model. It has been observed that energy time histories of a 4-DOF reduced model nearly coincided with target values.

When dealing with accuracy versus sped-up, it has been shown that the POD can decrease the number of DOFs from the original 39 (one at each storey) to just 1, guaranteeing an accuracy of 0.99 (1 being featured by the original model) according to what is explained in this study, and leading to a speedup in the computations higher than 500. We have also shown that, to further match higher order frequency oscillations (accuracy of 0.99999), the retained degrees of freedom result to be increased to 4, still obtaining a speedup higher than 200.

It has been shown that the POD based reduced models are robust to a change of loading as well; moreover, the models built by snapshots resulting from simulations of the full model subject to El Centro record feature the same level of accuracy when are shaken by Kobe and Friuli record.

In following chapters, the reduced model built by POD will be incorporated into Bayesian filters to assess the capabilities of such an approach in state estimation of non-damaging and dual estimation of damaging structures, possibly detecting and locating the occurring damage.

References

Al-Dmour AS, Mohammad KS (2002) Active control of flexible structures using principal component analysis in the time domain. J Sound Vib 253:545–569

Aschheim MA, Black EF, Cuesta I (2002) Theory of principal components analysis and applications to multistory frame buildings responding to seismic excitation. Eng Struct 24:1091–1103

Barbella G, Perotti F, Simoncini V (2011) Block Krylov subspace methods for the computation of structural response to turbulent wind. Comput Methods Appl Mech Eng 200:2067–2082

Buljak V (2012) Inverse analyses with model reduction: proper orthogonal decomposition in structural mechanics. Springer, New York

Buljak V, Maier G (2011) Proper orthogonal decomposition and radial basis functions in material characterization based on instrumented indentation. Eng Struct 33:492–501

De Boe P, Golinval J (2003) Principal component analysis of a piezosensor array for damage localization. Struct Health Monit 2:137–144

Eftekhar Azam S, Mariani S (2013) Investigation of computational and accuracy issues in POD-based reduced order modeling of dynamic structural systems. Eng Struct 54:150–167

Feeny BF (2002) On proper orthogonal co-ordinates as indicators of modal activity. J Sound Vib 255:805–817

Feeny BF, Kappagantu R (1998) On the physical interpretation of proper orthogonal modes in vibrations. J Sound Vib 211:607–616

Fukunaga K (1990) Introduction to statistical pattern recognition. Academic Press Inc, London

Galvanetto U, Surace C, Tassotti A (2008) Structural damage detection based on proper orthogonal decomposition: experimental verification. AIAA J 1624–1630

Georgiou I (2005) Advanced proper orthogonal decomposition tools: using reduced order models to identify normal modes of vibration and slow invariant manifolds in the dynamics of planar nonlinear rods. Nonlinear Dyn 41:69–110

Gutiérrez E, Zaldivar JM (2000) The application of Karhunen-Loève, or principal component analysis method, to study the non-linear seismic response of structures. Earthquake Eng Struct Dynam 29:1261–1286

Han S, Feeny B (2003) Application of proper orthogonal decomposition to structural vibration analysis. Mech Syst Signal Process 17:989–1001

Hemez FM, Doebling SW (2001) Review and assessment of model updating for non-linear, transient dynamics. Mech Syst Signal Process 15:45–74

Jolliffe IT (1986) Principal component analysis. Springer, New York

Karhunen K (1947) Uber lineare methoden in der wahrscheinlichkeitsrechnung. Annales Academiae Scientiarum Fennicae, Series A1: Mathematica-Physica, vol 37, pp 3–79

Kerschen G, Golinval JC (2002) Physical interpretation of the proper orthogonal modes using the singular value decomposition. J Sound Vib 249:849–865

Kerschen G, Golinval J, Vakakis A, Bergman L (2005) The method of proper orthogonal decomposition for dynamical characterization and order reduction of mechanical systems: an overview. Nonlinear Dyn 41:147–169

Klema VC, Laub AJ (1980) Singular value decomposition: its computation and some applications. IEEE Trans Autom Control AC-25:164–176

Kosambi D (1943) Statistics in function space. J Indian Math Soc 7:76–88

Lenaerts V, Kerschen G, Golinval J (2003) Identification of a continuous structure with a geometrical non-linearity. Part II: proper orthogonal decomposition. J Sound Vib 262:907–919

Liang YC, Lee HP, Lim SP, Lin WZ, Lee KH, Wu CG (2002a) Proper orthogonal decomposition and its applications—Part I: theory. J Sound Vib 252:527–544

Liang YC, Lin WZ, Lee HP, Lim SP, Lee KH, Sun H (2002b) Proper orthogonal decomposition and its applications—Part II: model reduction for MEMS dynamical analysis. J Sound Vib 256:515–532

Lucia DJ, Beran PS, Silva WA (2004) Reduced-order modeling: new approaches for computational physics. Prog Aerosp Sci 40:51–117

Mees AI, Rapp PE, Jennings LS (1978) Singular-value decomposition and embedding dimension. Phys Rev 36:340–346

North GR (1984) Empirical orthogonal functions and normal modes. J Atmos Sci 41:879–887

Obukhov AM (1954) Statistical description of continuous fields. T Geophys Int Akad Nauk USSR 24:3–42

Park S, Lee J, Yun C, Inman DJ (2008) Electro-mechanical impedance-based wireless structural health monitoring using PCA-data compression and k-means clustering algorithms. J Intell Mater Syst Struct 19:509–520

Pearson K (1901) On lines and planes of closest fit to systems of points in space. Phil Mag 2:559–572

Preisendorfer RW (1979) Principal components and the motions of simple dynamical systems. Scripps Institution of Oceanography

Samadiani E, Joshi Y (2010) Reduced order thermal modeling of data centers via proper orthogonal decomposition: a review. Int J Numer Meth Heat Fluid Flow 20:529–550

Schilders W (2008) Introduction to model order reduction. In: Heres P, Schilders W (eds) Model order reduction: theory, research aspects and applications, Mathematics in Industry, pp 3–32

Shane C, Jha R (2011) Proper orthogonal decomposition based algorithm for detecting damage location and severity in composite beams. Mech Syst Signal Process 25:1062–1072

Sirovich L (1987) Turbulence and the dynamics of coherent structures. Q Appl Math 45:561–571

Smith TR, Moehlis J, Holmes P (2005) Low-dimensional modelling of turbulence using the proper orthogonal decomposition: a tutorial. Nonlinear Dyn 41:275–307

Steindl A, Troger H (2001) Methods for dimension reduction and their application in nonlinear dynamics. Int J Solids Struct 38:2131–2147

Tadmor G, Noack BR, Morzyński M (2006) Control oriented models and feedback design in fluid flow systems: a review. In: 14th Mediterranean conference on control and automation, MED'06

Thomas JP, Dowell EH, Hall KC (2003) Three-dimensional transonic aeroelasticity using proper orthogonal decomposition-based reduced-order models. J Aircr 40:544–551

Tubino F, Carassale L, Solari G (2003) Seismic response of multi-supported structures by proper orthogonal decomposition. Earthquake Eng Struct Dynam 32:1639–1654

Yadalam VK, Feeny BF (2011) Reduced mass-weighted proper decomposition for modal analysis. J Vib Acoust Trans ASME 133

Chapter 4
POD-Kalman Observer for Linear Time Invariant Dynamic Systems

Abstract This Chapter investigates the statistical properties of residual errors induced by POD-based reduced order modeling. Such errors enter into the state space equations of the reduced systems in terms of system evolution and observation noise. A fundamental assumption made by recursive Bayesian filters, as exploited in this study, is the whiteness of the aforementioned noises. In this chapter, null hypothesis of the whiteness of the noise signals is tested by making use of the Bartlett's whiteness test. It is shown that, no matter what the number of POMs retained in the analysis is, the null hypothesis of the whiteness is always to be rejected. However, the spectral power of the embedded periodic signals decreases rapidly by increasing the number of POMs. The speed-up gained by incorporating POD-based reduced models into Kalman observer of linear time invariant systems, is also addressed in this chapter. It is shown that the reduced models incorporated into the Kalman filter dramatically reduce the computing time, leading to speed-up of 300 for a POD model featuring 1 POM, which is able to accurately reconstruct the displacement time history of the structure. Moreover, it is revealed that the coupling of POD and Kalman filter can improve the estimations provided by POD alone.

4.1 Introduction

To develop an online and real-time algorithm for the detection of damage in structural systems is the ultimate goal of this monograph project. To accomplish this goal, we have primarily studied the possibility of exploiting Bayesian filters to fulfill the objective of this study in Chap. 2. However in the case of multi-storey buildings, it was revealed that as the number of floors increases, the bias in the estimation of parameters as well as in damage detection increases.

Subsequently, we propose to use reduced order models in combination with Bayesian filters to monitor the state of the structure. Moreover, the efficiency of POD in terms of speed-up and accuracy has been examined numerically in the

S. Eftekhar Azam, *Online Damage Detection in Structural Systems*,
PoliMI SpringerBriefs, DOI: 10.1007/978-3-319-02559-9_4,
© The Author(s) 2014

previous Chapter. On the other hand, this chapter considers the numerical assessment of the efficiency of POD-based reduced order models in state estimation of linear time-invariant structural systems. The Kalman filter as a popular means provides the optimal estimates of the state of a linear state-space model affected by white Gaussian noises. Nevertheless next, we will illustrate that the uncertainties induced by the POD are not white noises.

The analysis of the linear time-invariant model allows us to analyze the effect of uncertainties induced by the POD on the optimal performance of the Kalman filter. Therefore, the reduced model of the system is incorporated into a Kalman filter; moreover by assuming that a minimal number of observables are managed, the speed-up and accuracy of state estimation is examined. As a well-known fact, the POD models are not strongly constructed to a change in the parameters of the system; in fact, proper orthogonal modes (POMs) were somehow employed as indicators of the damage in various structures, such as beams (Galvanetto and Violaris 2007a), trusses (Ruotolo and Surace 1999) and composite materials (Shane and Jha 2011a). If the system is exposed to unpredictable change in the parameter, e.g. due to inception or growth of damage, the reduced model fails to be accurate. However, a potential application of an approximated linear time invariant model in automatic control of the structural response (Gustafsson and Mäkilä 1996) encourages the search for high fidelity and computationally efficient reduced models. To estimate the state of a system, even in an accurate approach, does not explicitly include information on the damage: therefore in the next Chapter, we will acknowledge damage detection via Bayesian filtering and reduced order modeling. In the sequel, the necessity of employing observers in structural feedback control is primarily discussed; subsequently, the statistical properties of the residual error process is evaluated in order to verify if they fulfill the requirements (whiteness and Gaussianity) of Kalman filter to provide the optimal solution. Hence, the Kalman-POD observer is concisely reconsidered. The Chapter is finally concluded by illustrating the performance of Kalman-POD observer: the efficiency of the algorithms is evaluated to ensure strength to change in the seismic excitation source, as it was performed in Chap. 3. Furthermore, the effect of correlated uncertainties in the performance of Kalman-POD observer is examined. The computational gain obtained by the application of Kalman-POD observer, when solely compared to Kalman observer, is presented in terms of speed-up gained in computations.

4.2 Structural Feedback Control and the Kalman Observer

Feedback control is aimed to be employed to develop automated algorithms in order to harness response of the systems (Goodwin et al. 2001). Clock regulating devices and mechanisms to keep wind-mills pointed toward the wind are the early instances of control systems. During industrial revolution, invention of machinery to transform raw materials into commodities, specifically steam engine, which includes transforming a large amount of energy to mechanical work, caused engineers to perceive the need for organized control strategies of the power consumed by machinery in order to guarantee the safe operation of the facilities (Goodwin et al. 2001). Currently, control engineering has become an omnipresent element of industry. Although industrial instances of feedback control date back to the nineteenth century, its applications are recently being employed in structural engineering field. Since 1990, automatic control strategies have gained popularity to further extend life cycle and performance of earthquake resistant structural systems. To reexamine the real applications of active structural control in Japan, one can study (Ikeda 2009) which the application of active tuned mass dampers for vibration suppression of high rise buildings exposed to lateral loads is discussed. To review a list concerning the active control strategies employed in other building types including bridges, tensegrity structures and trusses, one can refer to Korkmaz (2011). Control algorithm design is accomplished by incorporating many disciplines of science and technology, including but not limited to modeling (to capture the underlying dynamics of the system), sensors (to measure state of the system), actuators (to force the system to follow the desired trajectory), communications (to transmit the data) and computations (to calculate action data based on measured observations) (Goodwin et al. 2001). This chapter of the monograph is devoted to develop computationally efficient reduced models for their possible application in control of seismically excited multi storey buildings.

Control algorithms will not be discussed. However, to explain how system control terms enter the state-space equations in further details and to describe the need for the models in structural control, a linear time invariant system is considered and its state-space equations is written as:

$$z_k = F z_{k-1} + G e_k + v_k \qquad (4.1)$$

$$y_k = H z_k + w_k \qquad (4.2)$$

where: z_k represents the state of the system (e.g. displacement, velocity and acceleration of each storey in a structure) at time instant t_k; e_k is the control input, which is computed by using control algorithms in order to restrict the state of structure to a desired reference; y_k denotes the noisy system observations; F maps the state over time; G links the control feedback to the relevant degrees-of-freedom and H links the observation and state; v_k and w_k are evolution and observation uncertainties. The idea in the state space approach to feedback control, is to synthesize a full state feedback through:

$$e_k = K z_k \qquad (4.3)$$

where K, the gain matrix, is computed to satisfy the objective of the closed loop system; in a civil structure such an objective will be, e.g. the suppression of vibrations induced by external loads (e.g. loads or seismic excitations).

The problem is that, in most practical cases, the state vector is not fully known: it may require too many sensors, or it may be due to technical reasons (for instance, displacements of a multi-storey structure are difficult to monitor).

The process of reconstructing the entire state of a system, based on a physical model and observation signals, is called observer design (Preumont 2011). It is known that, dealing with linear state-space models, provided that the distribution of the uncertainties is Gaussian and there is no correlation in uncertainty time series, Kalman Filter furnishes the optimal observer of the system (Preumont 2011). This chapter of the monograph deals with the reduction of the computational cost of a Kalman observer of the linear time invariant dynamic systems, by utilizing a surrogate POD-based reduced model of the system to be incorporated into the Kalman filter algorithm. The efficiency of POD for model reduction of models studied in current Chapter, in terms of speed-up and accuracy, has been ascertained in Chap. 3, where it has been shown that the POD can be a reasonable candidate to reduce the computational costs of structural analysis.

4.3 Statistical Assessment of Residual Errors Induced by POD

We start by recalling from Chap. 3 the set of ordinary differential equation that governs the dynamics of a structural system:

$$M\ddot{u} + Ku = R(t) \tag{4.4}$$

where: M and K are the stationary mass and stiffness matrices, respectively; $R(t)$ is the external load vector; $\ddot{u}$ and u are the storey acceleration and displacement vectors, respectively.

By utilizing a Newmark time-integration algorithm, (4.4) is discretized in the time domain, through definition of the vector $z_k = \begin{bmatrix} u_k & \dot{u}_k & \ddot{u}_k \end{bmatrix}^T$ at time t_k. The discrete state space form of (4.4) reads:

$$z_k = A z_{k-1} + B_k + v_k \tag{4.5}$$

$$y_k = H z_k + w_k \tag{4.6}$$

where:

$$A = \begin{bmatrix} I - \beta \Delta t^2 M^{-1} K & \Delta t I - \beta \Delta t^3 M^{-1} K & -\beta\left(\frac{1}{2} - \beta\right)\Delta t^4 M^{-1} K + \Delta t^2\left(\frac{1}{2} - \beta\right)I \\ -\gamma \Delta t M^{-1} K & I - \gamma \Delta t^2 M^{-1} K & -\gamma\left(\frac{1}{2} - \beta\right)\Delta t^3 M^{-1} K + \Delta t(1 - \gamma)I \\ -M^{-1} K & -\Delta t M^{-1} K & -\Delta t^2\left(\frac{1}{2} - \beta\right)M^{-1} K \end{bmatrix} \tag{4.7}$$

and:

$$B_k = \begin{bmatrix} \beta \Delta t^2 M^{-1} R_k \\ \gamma \Delta t M^{-1} R_k \\ M^{-1} R_k \end{bmatrix} \tag{4.8}$$

v_k and w_k are evolution and measurement noises, assuming the full model to be deterministic, former one is not considered to enter the evolution of state of the system, while latter is assumed to be a stationary zero mean white Gaussian noise featuring time invariant covariance matrix of W.

With the same notation of Chap. 3, the reduced order model of the system can now be written as:

$$M_l \ddot{\alpha}(t) + K_l \alpha(t) = R_l(t) \tag{4.9}$$

where: α is the coordinate of the reduced model and governs the evolution in time of the structural response along the POMs. Once the solution of (4.9) is obtained, the full state of the system can be computed by making use of (4.9):

$$\ddot{u} \approx \Phi_l \ddot{\alpha} \quad \dot{u} \approx \Phi_l \dot{\alpha} \quad u \approx \Phi_l \alpha \tag{4.10}$$

or equivalently:

$$\begin{Bmatrix} u \\ \dot{u} \\ \ddot{u} \end{Bmatrix} \approx \begin{bmatrix} \Phi_l & 0 & 0 \\ 0 & \Phi_l & 0 \\ 0 & 0 & \Phi_l \end{bmatrix} \begin{Bmatrix} \alpha \\ \dot{\alpha} \\ \ddot{\alpha} \end{Bmatrix} = L \begin{Bmatrix} \alpha \\ \dot{\alpha} \\ \dot{\alpha} \end{Bmatrix}. \tag{4.11}$$

Hence, the reduced state space model of the system can be obtained by coupling the time evolution of the coordinates of the reduced model and the observation equation. By definition of the vector $z_{r,k} = [\alpha_k \quad \dot{\alpha}_k \quad \ddot{\alpha}_k]^T$, the state space equation reads:

i.e.:

$$z_{r,k} = A_r z_{r,k-1} + B_{r,k} + v_k \tag{4.12}$$

$$y_k = HLz_{r,k} + w_k \tag{4.13}$$

where:

$$A_r = \begin{bmatrix} I - \beta \Delta t^2 M_l^{-1} K_l & \Delta t I - \beta \Delta t^3 M_l^{-1} K_l & -\beta(1/2 - \beta)\Delta t^4 M_l^{-1} K_l + \Delta t^2(1/2 - \beta)I \\ -\gamma \Delta t M_l^{-1} K_l & I - \gamma \Delta t^2 M_l^{-1} K_l & -\gamma(1/2 - \beta)\Delta t^3 M_l^{-1} K_l + \Delta t(1 - \gamma)I \\ -M_l^{-1} K_l & -\Delta t M_l^{-1} K_l & -\Delta t^2(1/2 - \beta)M_l^{-1} K_l \end{bmatrix} \tag{4.14}$$

and:

$$B_{r,k} = \begin{bmatrix} \beta \Delta t^2 M_l^{-1} R_{l,k} \\ \gamma \Delta t M_l^{-1} R_{l,k} \\ M_l^{-1} R_{l,k} \end{bmatrix}. \tag{4.15}$$

Since it is assumed that the original model is deterministic, v_k is solely attributed to inaccuracy of the reduced model; w_k instead is representative of measurement errors and model reduction inaccuracies together. In case v_k and w_k are white Gaussian noises, the Kalman filter can furnish the optimal estimates of the state of the reduced model; on the contrary, if the distributions of the uncertainties are not Gaussian, uncorrelated or a combination thereof, the performance of Kalman filter is not a priori known to be satisfactory.

In this section, Bartlett white noise test (Bartlett 1978) is profited to verify the null hypothesis of whiteness of the errors induced by the reduced order modeling. In this regard, Bartlett test compares the empirical cumulative normalized periodogram of the given signal with the cumulative distribution of a uniform random variable. The periodogram of an arbitrary random signal (e.g. $s_k, k = 1, 2, \ldots, N$), as a mean for spectral analysis, is defined as (Stoica and Moses 1997):

$$I(\omega) = \frac{1}{N} \left| \sum_{k=1}^{N} s_k e^{-i\omega k} \right| \tag{4.16}$$

while, the cumulative periodogram is computed:

$$J(\omega_k) = \frac{\sum_{i=1}^{k} I(\omega_i)}{\sum_{j=1}^{N} I(\omega_j)}. \tag{4.17}$$

To perform the comparison, and measure the possible deviation from the whiteness assumption, the Kolmogorov–Smirnov statistics is adopted by Bartlett test (Reschenhofer 1989). In case the associated Kolmogorov–Smirnov statistics of the test exceeds the critical values, for a given confidence interval, the null hypothesis of whiteness will be rejected. For each sample size, and for several confidence levels, the critical values of Kolmogorov–Smirnov statistics are tabulated and reported in references Miller (1956), Kececioglu (2002). The highest confidence interval, for which the test statistics are reported in Kececioglu (2002), are related to a probability equal to 99 %; therefore, to accept or reject the hypothesis by the maximum probability, in this chapter, we compare test statistics to the value associated with probability of 99 %. The critical values of the test statistics also depend on the sample size, which in our case is the length of the error signal. These critical values are estimated trough Monte Carlo simulations (Lilliefors 1967): if the sample size (N) is higher than 35, the critical value of the test statistics is curve fitted and is represented by $\frac{1.63}{\sqrt{N}}$ (Kececioglu 2002). It is reported that the Bartlett test is not a suitable method to test whiteness of observation signals with small sample sets (Reschenhofer 1989). However, dealing with time series of error signal, there is practically no limitation in increasing the number of the samples, and samples size issues are not affecting the test results. The results of the test are reported graphically, where empirical cumulative normalized periodogram of the given signal and the cumulative distribution of a uniform random variable (a straight line, passing from the origin and with a slope equal to the inverse of the Nyquist frequency), accompanied by two lines representing the confidence interval, are plotted in the same graph.

Assuming that the true dynamics of the system is known and obtained by analysis of the full model, the errors induced by the model order reduction are defined as the difference between the true dynamics of the system and the dynamics furnished by the reduced model in this study. The error is considered in terms of difference between the physical coordinates (i.e. u, $\dot{u}$, $\ddot{u}$) and the POD temporal coordinates (i.e. α, $\dot{\alpha}$, $\ddot{\alpha}$). At time instant t_k, the error signals can therefore be written as:

$$o_k = \ddot{u}_k - \Phi_l \ddot{\alpha}_k \quad p_k = \dot{u}_k - \Phi_l \dot{\alpha}_k \quad q_k = u_k - \Phi_l \alpha_k \tag{4.18}$$

while the errors concerning POD coordinates are:

$$\rho_k = \Phi_l^T \ddot{u}_k - \ddot{\alpha}_k \quad \sigma_k = \Phi_l^T \dot{u}_k - \dot{\alpha}_k \quad \tau_k = \Phi_l^T u_k - \alpha_k. \tag{4.19}$$

It is observed in (4.18) and (4.19) that the error signals relevant to velocity and acceleration are not assumed as temporal derivatives of displacement error signal. This fact is due to the uncertainties induced by the model order reduction.

In the next section, it is shown that the errors in the reconstructing the state of the full model affects the observation equation of the reduced state space model. Instead the error in the reconstructing the state of the reduced model enters and affects the evolution equation of the reduced model.

4.4 Formulation of Kalman-POD Observer for Linear Time Invariant Systems

The bulk of Chap. 2 has been dedicated to Bayesian filters for the estimation of states and parameters of mechanical systems, of which only a part of the state is observed. However, to keep this chapter self-contained, key points of recursive Bayesian estimation of mechanical systems are reviewed. The outline of all the Bayesian filters can be drawn in the two stages of prediction and update: in the prediction stage, a model of the system is employed to predict the dynamics of the entire state vector, whereas in the update stage, as observations from a part of the state, or as measurable quantities which are correlated with the state become available, the entire state vector is updated. For instance, in a multi storey building, it is expensive or even practically impossible to measure displacements of the storeys directly, while accelerations are easy to measure. In such cases, provided that a model of the structure is available and the model is linear, if uncertainties in the model and in the measurements are uncorrelated Gaussian noises the Kalman filter is the optimal tool to estimate the state of the system.

In practice, it may happen that the high dimension of the model of the structure prevent the filter to fulfill its task in real-time. In such a case, exploiting a reduced model will be beneficial for reducing the computational cost of the Kalman filter.

Table 4.1 POD-Kalman observer

- Initialization at time t_0:

$$\widehat{z}_{r,0} = \mathbf{\Phi}_l^T \mathbb{E}[z_0]$$
$$\mathbf{P}_{r,0} = \mathbf{\Phi}_l^T \mathbb{E}\left[(z_0 - \widehat{z}_0)(z_0 - \widehat{z}_0)^T\right]\mathbf{\Phi}_l$$

- At time t_k, for $k = 1, \ldots, N_t$:
- Prediction stage:
  1. Evolution of state and prediction of covariance

$$z_{r,k}^- = A_{r,k}z_{r,k-1} + B_{r,k}$$
$$P_{r,k}^- = A_{r,k}P_{r,k-1}A_{r,k}^T + V$$

- Update stage:
  1. Calculation of Kalman gain:

$$G_k = P_{r,k}^- L^{\mathrm{T}}H_{r,k}^{\mathrm{T}}\left(H_{r,k}L P_{r,k}^- L^{\mathrm{T}}H_{r,k}^{\mathrm{T}} + W\right)^{-1}$$

  2. Improve predictions using latest observation:

$$\widehat{z}_{r,k} = z_{r,k}^- + G_k\left(y_k - H_{r,k}L z_{r,k}^-\right)$$
$$P_{r,k} = P_{r,k}^- - G_k H_{r,k}L P_{r,k}^-$$

- Reconstruction stage:

$$\widehat{z}_k = L\widehat{z}_{r,k}$$

In this chapter, reduced models which are built by the POD are used to speed-up the calculations.

The idea of speeding up the calculations required by Kalman filters via reduced order modeling has been already exploited in meteorology in order to predict the near surface winds over the tropical Pacific ocean (Wikle and Cressie 1999). A set of empirical functions was adopted to reduce the computational burden of the reconstruction of the wind velocity field, via data available from a few observation points. Malmberg et al. (2005) adopted subspace realized by PCA to tackle the same problem. They assumed that the weather condition can be thought of as a linear combination of some dominant modes (the weather condition is modeled by a linear time invariant state-space model), the modes being supposed to be invariant; however, the contribution of each mode may vary over time, and the Kalman filter was used to estimate the contribution of each one. Though the concept of reduced state-space Kalman filter is gaining popularity in meteorology (He et al. 2011; Tian et al. 2011), its possible application has not been considered in structural engineering field yet. In this section, we deal with the application of Kalman filter to estimate the POD coordinates of Eq. (4.12). At each time instant, after the reduced states are estimated, the entire state vector is reconstructed. For details concerning the synergy of POD and Kalman filter, see Table 4.1.

Provided that the reduced model of the structure is already available, it is observed that the algorithm is simply the application of a Kalman filter to estimate the current state of a linear time-invariant system. In such a system, a linear combination of POMs can represent the dynamics of the system. The POMs are constant over time and do not change; however, the contribution of each mode in the construction of the response of the structure is changing over time.

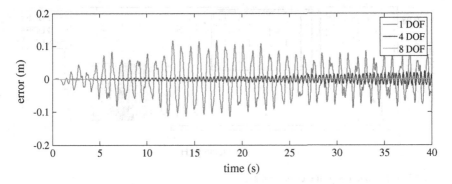

Fig. 4.1 Errors in the displacement time histories furnished by reduced order models

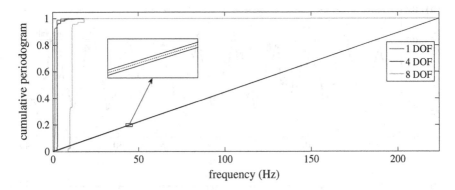

Fig. 4.2 Cumulative periodograms of error signals

The Kalman filter, based on the observation made from a part of state vector (e.g. accelerations of some storeys) quantifies the contribution of each POM in the estimation of the state of the system.

4.5 Numerical Assessment of POD-Kalman Observer for Seismic Analysis of Linear Time Invariant Systems

As a case study, in Chap. 3 we investigated the capability of POD in speeding up the computations required to model the dynamics of the Pirelli Tower in Milan; in this Section, whiteness of the uncertainties in the reduced models built in Chap. 3 is primarily assessed, so as to verify that the requirements of the Kalman filer for optimal performance are satisfied. Subsequently, robustness of the Kalman-POD approach to changes in the seismic excitation source is investigated. The Section finally concludes with the numerical assessment of speed-up and accuracy of the Kalman-POD algorithm.

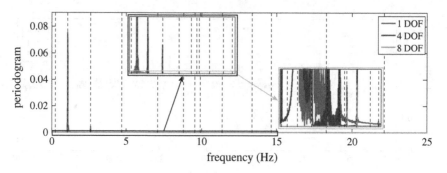

Fig. 4.3 Periodograms of the error signals

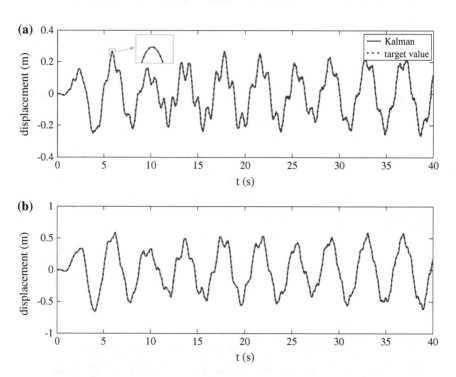

Fig. 4.4 Time histories of the horizontal displacements of 20th floor (**a**) and 39th floor (**b**) as induced by the El Centro earthquake, performance of the Kalman filter

As for the error of reduced models for reconstructing the displacement history of the roof floor, Fig. 4.1 shows the relevant error for reduced models with various number of retained POMs. The errors are related to the analysis of the building when acceleration time history of E1 Centro earthquake is applied to shake the structure. It is observed that, by increasing the number of POMs, the amplitude of the error signal drastically decreases. However, from the time evolution of the error

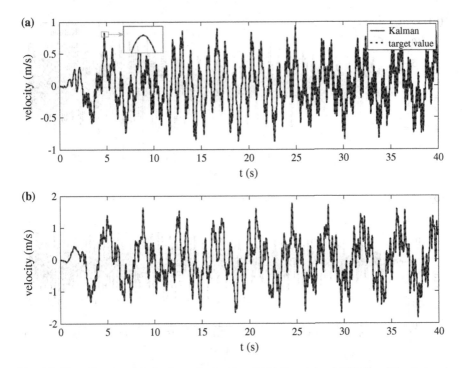

Fig. 4.5 Time histories of the horizontal velocities of 20th floor (**a**) and 39th floor (**b**) as induced by the El Centro earthquake, performance of the Kalman filter

signals relevant to reduced models featuring different number of POMs, it seems that there is a strong correlation in them, as the signals look similar to a sinusoid with a time varying amplitude. This is corroborated by the cumulative periodograms of the error signals shown in Fig. 4.2. By increasing the number of POMs retained in the reduced models from one to eight, despite the decrease in the error amplitudes, the hypothesis of the whiteness can still be rejected, as all three periodograms relevant to the reduced model exceed the 99 % confidence interval (indicated by two parallel black lines in the closeup as presented in Fig. 4.2). By examining at the cumulative periodograms, it can be observed that, as the number of POMs of the reduced model increases, the main jumps move to higher frequency zones.

To investigate this issue in further details, we examine the periodograms of the error signals as displayed in Fig. 4.3. To facilitate the comparison, the first few natural frequencies of the structure are indicated by vertical dashed lines (see Table 3.1). It is observed that the main peak in the error of the 1-DOF reduced model is coincident in the second natural frequency of the structure. By increasing the number of DOFs of the reduced models, according to the reduction in the error amplitude already illustrated in Fig. 4.3, the power of the harmonic components embedded in the signal attenuates severely to the extent that it is not possible to distinguish the corresponding peaks in Fig. 4.3.

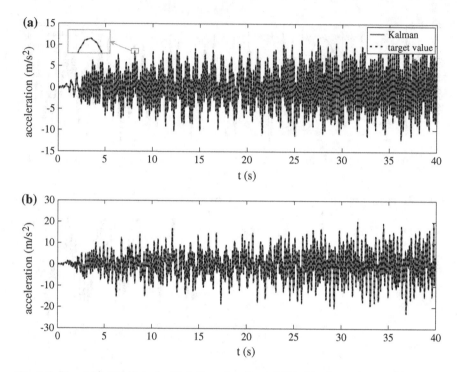

Fig. 4.6 Time histories of the horizontal accelerations of 20th floor (**a**'0 and 39th floor (**b**) as induced by the El Centro earthquake, performance of the Kalman filter

Close-ups in Fig. 4.3 allow us to compare the spectral power of the error of three reduced models in a clearer fashion: it is observed that the main periodicity of the error signal of the 4-DOF reduced model coincides with third natural frequency of the system; moreover, the close up further indicates that, in frequency content of the error signal of the 8-DOF reduced model, the first peak is coincident in the 8th natural frequency of the system. The trend in Fig. 4.1 suggests that as the number of DOFs of the reduced model increases, the amplitude of the error signal decreases; consequently, the spectral power of the error signal decreases as well. In addition, as the number of DOFs retained in the reduced model increases, the dominant frequency contents coincide with higher natural frequencies of the system. This trend suggests that the subspace spanned by POMs has a degree of similarity with the subspace spanned by the eigenmodes of the system: frequency content of the error induced by neglected POMs is coincident in the higher order eigen-frequencies of the structure.

In what precedes, it was observed that the uncertainties in the errors of reduced order models are correlated, and not white noises; hence, the optimal performance of the Kalman observer is not guaranteed. However, it was further indicated that,

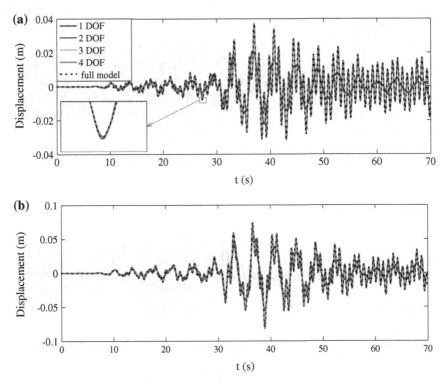

Fig. 4.7 Time histories of the horizontal displacements of 20th floor (**a**) and 39th floor (**b**) as induced by the Friuli earthquake, performance of the POD-Kalman

by an increase in the number of POMs retained in the reduced model, the spectral density of the correlation in the errors diminishes rapidly.

In proceeding section, the performance of the Kalman observer, if applied to the estimation of the entire state vector on the basis of observations of the acceleration time history of the 39th storey (roof floor) is assessed. To choose other storeys for observation, or add more data yields the similar results: it is known that state of a linear state space model with white Gaussian noises is optimality estimated through the Kalman observer. In Figs. 4.4, 4.5 and 4.6, displacement, velocity and acceleration time histories of the 20th (mid floor) and 39th (roof) floors are illustrated as representative outcomes for the performance of the filter.

In the analysis for numerical assessment of performance of the Kalman filter, the evolution equation is assumed to be deterministic, and the noise in the observations is supposed to be a white stationary Gaussian process. As expected from optimality of the Kalman observer for dealing with aforementioned problems, it is observed that the estimates furnished by the Kalman filter nearly coincide with the target values. This fact is observed through the close-ups in each time history graph.

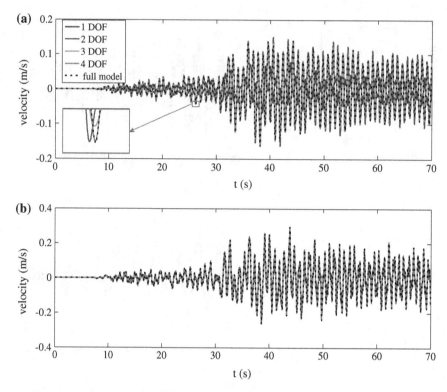

Fig. 4.8 Time histories of the horizontal displacements of 20th floor (**a**) and 39th floor (**b**) as induced by the Friuli earthquake, performance of the Kalman-POD

In the remainder of this Section, the performance of Kalman-POD algorithm is assessed in order to estimate the state of the Pirelli tower. As it has been illustrated, the uncertainties in the state-space model are not white; consequently, the performance of the Kalman observer is not a priori known. In this chapter, we utilize the POD-based reduced models; thus the readers are referred to see Chap. 3 for the details. The reduced model is used by snapshots taken from the simulation of the response of the full model to El Centro accelerogram excitation. Figures 4.7, 4.8 and 4.9 demonstrate time histories of the estimations of displacements, velocities and accelerations of 20th and 39th floor via Kalman-POD algorithm, when the building is shaken by Friuli acceleration record. It is observed that by keeping only 3 POMs in the reduced model, the time histories estimated by POD-Kalman match those of the full model. To have insights on the improvement in the quality of the estimates by Kalman-POD when it is compared to POD, Tables 4.2 and 4.3 report the residual mean squared error (RMSE) of the 20th and 39th floors, respectively.

In Table 4.2 it is observed that, as the number of DOFs in the reduced model increases, the RMSE error of reconstruction of displacements, velocities and accelerations realized by the POD rapidly decreases. When using reduced models

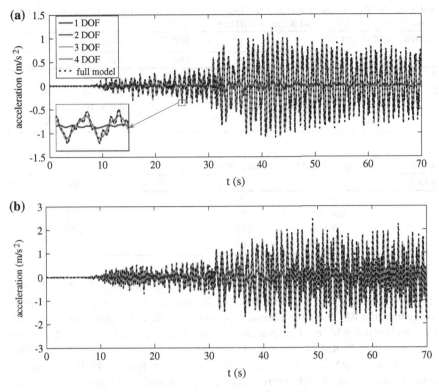

Fig. 4.9 Time histories of the horizontal accelerations of 20th floor (**a**) and 39th floor (**b**) as induced by the Friuli earthquake, performance of the POD-Kalman

Table 4.2 RMSE of time histories of displacements, velocities and accelerations of 20th floor, comparison between POD and Kalman-POD approaches

#DOFs	RMSE of POD			RMSE of POD-Kalman		
	Disp.	Vel.	Acc.	Disp.	Vel.	Acc.
1	5.9×10^{-5}	2.7×10^{-3}	1.4×10^{-1}	9.6×10^{-5}	4.4×10^{-3}	3.3×10^{-1}
2	1.0×10^{-6}	1.0×10^{-4}	1.2×10^{-2}	7.0×10^{-7}	1.9×10^{-4}	1.7×10^{-1}
3	2.0×10^{-8}	1.5×10^{-5}	1.4×10^{-2}	1.7×10^{-8}	1.1×10^{-5}	1.1×10^{-2}
4	5.0×10^{-9}	2.0×10^{-6}	2.7×10^{-3}	1.3×10^{-9}	1.3×10^{-6}	1.8×10^{-3}

with 1 and 2 POMs, the RMSEs of POD solely are less than those of Kalman-POD. However, moving to reduced models with 3 and 4 POMs Kalman-POD is able to improve the quality of the estimate with respect to what the application of POD alone offers. This phenomenon is mainly due to the high spectral power of the correlation structure embedded in the error signal: it has been revealed that by increasing the POMs retained in the reduced model, the spectral power of the noise correlations decrease rapidly.

Table 4.3 RMSE of time histories of displacements, velocities and accelerations of 39th floor, comparison between POD and Kalman-POD approaches

#DOFs	RMSE of POD			RMSE of POD-Kalman		
	Disp.	Vel.	Acc.	Disp.	Vel.	Acc.
1	1.1×10^{-4}	5.7×10^{-3}	4.6×10^{-1}	2.6×10^{-5}	1.3×10^{-3}	1.2×10^{-5}
2	5.0×10^{-6}	1.0×10^{-3}	3.1×10^{-1}	1.9×10^{-6}	1.6×10^{-4}	1.8×10^{-6}
3	3.9×10^{-6}	9.0×10^{-40}	2.4×10^{-1}	1.7×10^{-8}	4.0×10^{-6}	1.1×10^{-6}
4	8.7×10^{-7}	2.4×10^{-4}	7.8×10^{-2}	6.8×10^{-9}	1.3×10^{-6}	9.8×10^{-7}

Table 4.4 Speed-up obtained by Kalman-POD and POD

# DOFs	p	Speed-up (Kalman-POD)	Speed-up (POD)
1	0.99	309	515
2	0.999	279	385
3	0.9999	225	276
4	0.99999	187	244

Moving to the 39th floor, whose acceleration is measured, it is observed that RMSE of accelerations estimated by POD-Kalman is several orders of magnitude lower than the RMSE of the estimates provided by the POD alone, see Table 4.3. Concerning the RMSE of displacements and velocities, it is observed that the estimates of POD-Kalman always are smaller than the estimates of POD. Unlike the 20th storey RMSEs, which the estimates of the Kalman-POD observer in several cases featured higher error when compared with POD alone, in this case, the RMSE of Kalman-POD always is lower compared to the POD. This is due to the fact that the response of the system is measured at 39th floor. The trend suggests that, as the number of POMs in the reduced model increases, the estimates of the POD-Kalman outperform the POD only.

Concerning the speedup obtained by reducing the order of the full model, similar to Chap. 3, in this study, the discussed results have been obtained with a personal computer featuring and Intel Core 2 Duo CPU E8400, with 4 GB of RAM, running Windows 7 x64 as operating system and performing the simulations with MATLAB version 7.6.0.324. The speedup values reported in Table 4.4 confirms the efficiency of Kalman-POD in reducing the computational costs related to the Kalman filter algorithm. It is observed that, using POD-based models incorporated in a Kalman observer can render the calculations hundreds of times faster.

4.6 Summary and Conclusion

In this section, the problem of monitoring the entire state of a structure via a numerical model and observations relevant to several points of interest is acknowledged. It has been illustrated that, dealing with a linear model of the Pirelli tower, when the building is shaken by the El Centro earthquake record, the Kalman filter can provide nearly perfect results by employing only acceleration time history of the last floor, as the observation signal.

Subsequently, the reduced models built via POD are introduced into the Kalman filter to reduce the computational cost of the filter. It has been revealed that the reduced models incorporated into the Kalman filter dramatically reduce the computing time, leading to speed-up of 300 for a POD model featuring 1 POM, which is able to accurately reconstruct the displacement time history of the structure. Moreover, it has been indicated that the coupling of POD and Kalman filter can improve the estimations provided by the POD alone.

This chapter has been limited to linear time invariant systems, the bulk of next Chapter will be instead dealing with the time-varying systems, when there is no a priori information concerning the variation of parameters.

References

Bartlett MS (1978) An introduction to stochastic processes with special reference to methods and applications. Cambridge University Press, London

Galvanetto U, Violaris G (2007) Numerical investigation of a new damage detection method based on proper orthogonal decomposition. Mech Sys Signal Process 21:1346–1361

Goodwin GC, Graebe SF, Salgado ME (2001) Control system design. Pearson, London

Gustafsson TK, Mäkilä PM (1996) Modelling of uncertain systems via linear programming. Automatica 32:319–334

He J, Sarma P, Durlofsky LJ (2011) Use of reduced-order models for improved data assimilation within an EnKF context. In: Proceedings of SPE reservoir simulation symposium 2011, vol. 2, pp 1181–1195

Ikeda Y (2009) Active and semi-active vibration control of buildings in Japan-practical applications and verification. Struct Control Health Monit 16:703–723

Kececioglu DB (2002) Reliability engineering handbook, vol 2. DEStech publications Inc, Pennsylvania

Korkmaz S (2011) A review of active structural control: Challenges for engineering informatics. Comput Struct 89:2113–2132

Lilliefors HW (1967) On the Kolmogorov-Smirnov test for normality with mean and variance. J Am Stat Assoc 62:399–402

Malmberg A, Holst U, Holst J (2005) Forecasting near-surface ocean winds with Kalman filter techniques. Ocean Eng 32:273–291

Miller LH (1956) Table of percentage points of Kolmogorov statistics. J Am Stat Assoc 51:111–121

Preumont A (2011) Vibration control of active structures: an introduction. Springer, Berlin

Reschenhofer E (1989) Adaptive test for white noise. Biometrika 76:629–632

Ruotolo R, Surace C (1999) Using SVD to detect damage in structures with different operational conditions. J Sound Vib 226:425–439

Shane C, Jha R (2011) Proper orthogonal decomposition based algorithm for detecting damage location and severity in composite beams. Mech Sys Signal Process 25:1062–1072

Stoica P, Moses RL (1997) Introduction to spectral analysis. Printice Hall Inc, Upper Saddle river

Tian X, Xie Z, Sun Q (2011) A POD-based ensemble four-dimensional variational assimilation method. Tellus, Ser A: Dyn Meteorol Oceanogr 63:805–816

Wikle CK, Cressie N (1999) A dimension-reduced approach to space-time Kalman filtering. Biometrika 86:815–829

Chapter 5
Dual Estimation and Reduced Order Modeling of Damaging Structures

Abstract In this chapter, the dual estimation and reduced order modeling of a damaging structure is studied. In this regard, proper orthogonal decomposition is considered for reduced order modeling in order to find a subspace which optimally captures the dynamics of the system. Through a Galerkin projection, the equations governing the dynamics of the system are projected onto the subspace provided by the proper orthogonal decomposition technique. It is proven that the subspace established by application of the proper orthogonal decomposition is sensitive to changes of the parameters; therefore, it can be profited in the algorithms for estimation of the damage incidence. As for the dual estimation goal, the extended Kalman filter and extended Kalman particle filter are adopted; both filters, in their so-called update stage, make a comparison between the latest observation and the prediction of the state of the system to quantify the required adjustment in the estimation of the state and parameters. In the case of the reduced order modeling, for realization of such a comparison, reconstruction of full state of the system is required, which is obviously possible only if the subspace is known. In this chapter, an adjustment of the dual estimation concept has led to an online estimation of the proper orthogonal modes, components of the reduced stiffness matrix and the states of the structure. This novelty can intuitively help to detect the damage in the structure, locate it and potentially identify its intensity.

5.1 Introduction

To detect changes in the mechanical properties of structural members, it can be assumed as a method to monitor their health. In many cases, to identify the damage in the structure, one can considered it as a reduction of the stiffness (Yang and Lin 2005). This may be caused due to failure of a member to sustain further action, or it can be due to degradation in its material properties. That means that damage detection in a structure can be modeled as a system identification problem. To deal with a linear structure, offline identification of system matrices can be carried out

S. Eftekhar Azam, *Online Damage Detection in Structural Systems*,
PoliMI SpringerBriefs, DOI: 10.1007/978-3-319-02559-9_5,
© The Author(s) 2014

via several robust algorithms; as for output only techniques, data driven stochastic subspace identification (SSI) algorithm is the de facto standard stochastic system identification method (Van Overschee and De Moor 1996), however, in recent years the research on developing new techniques e.g. blind source separation (Abazarsa et al. 2013a, b; Ghahari et al. 2013c). The aforementioned method is successfully applied to identify the modal parameters of multi-storey buildings (Ghahari et al. 2013a, b) and modal identification of long span bridges (Ghahari et al. 2013d). Moreover, subspace identification algorithm is instead extensively applied to identify deterministic input–output systems (Loh et al. 2011). The aforementioned methodologies include singular value decomposition (SVD) and QR decomposition techniques (Moaveni et al. 2011). Extension of such methodologies to online system identification is normally perceived via setting a fixed length moving time window; as new observations become available, new subspace identification is perceived. Computational costs associated with SVD and QR prevent real-time application of such methods. Several methods were proposed to reduce the computational burden of the SVD and QR operations based on updating SVD and QR decomposed matrices; moreover, they are made suitable for near real-time applications (Loh et al. 2011). In this study, damage detection has been tackled via dual estimation of state and stiffness parameters by utilizing recursive Bayesian filters in an online means. We have shown in Chap. 2 that, as the number of DOFs of the space model of the structure increases, biases frequently affect the estimates furnished by the filters. To manage this problem, dual estimation of state and parameters of a reduced model of the structure are employed as the last resort.

Nevertheless, dissimilar to the identification of the full model of the system, to estimate components of the reduced stiffness yield no precise information concerning the intensity and location of the damage. It is a well-known fact that appropriate orthogonal modes of the structures include information regarding location and intensity of the damage (Ruotolo and Surace 1999; Vanlanduit et al. 2005; Galvanetto and Violaris 2007; Shane and Jha 2011). Hence, this feature of POMs can potentially resolve deficiencies of parameter estimation of a reduced model as an indicator of damage location and severity. To accomplish this objective, an algorithm for dual estimation of state and parameters of a reduced model, accompanied by an online estimation of the POMs of the structure is suggested. The proposed procedure utilizes appropriate orthogonal decomposition for model order reduction; afterwards, it exploits Bayesian filters for dual estimation of the full state and reduced parameters of the system. At each recursion, Kalman filter is adopted to update the subspace spanned by the POMs retained in the reduced model. This method can effectively detect, locate and identify the severity of the damage in shear building type structures. The efficiency of the methodology is testified through pseudo experimental data obtained by employing direct analyses.

The proceeding sections of this chapter are organized as follows. In Sect. 5.2 the state space formulation of shear buildings is reexamined; moreover, key features of the reduced order state space model of the system are highlighted in Sect. 5.3. In Sect. 5.4 the peculiarities of dual estimation and reduced order modeling of a

Fig. 5.1 Schematic view of a shear building

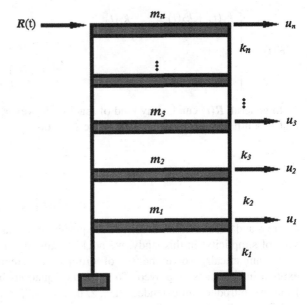

damaging structure are presented and discussed, and we define our proposal as how to tackle the problem. Finally, efficiency of our proposed method is numerically testified in Sect. 5.5.

5.2 State Space Formulation of Shear Building-Type Structural Systems

In this study, it is aimed to develop an algorithm for multi-storey buildings and to investigate shear buildings, i.e. models obtained by lumped mass assumption for each story, see Fig. 5.1.

Representing storey displacements, velocities and accelerations by u, $\dot{u}$ and $\ddot{u}$ respectively, the governing equation of motion of the building reads:

$$M\ddot{u} + D\dot{u} + K(t)u = R(t) \tag{5.1}$$

where M is the stationary mass matrix, D denotes time invariant damping matrix and $K(t)$ stands for time varying stiffness matrix, whose variation in time is due to possible damage phenomena and is usually unpredictable; $R(t)$ is the loading vector:

$$M = \begin{bmatrix} m_1 & & & \\ & m_2 & & \\ & & \ddots & \\ & & & m_n \end{bmatrix} \tag{5.2}$$

$$
K(t) = \begin{bmatrix} k_1(t) + k_2(t) & -k_2(t) \\ -k_2(t) & k_2(t) + k_3(t) \\ & & \ddots \\ & & & k_{n-1}(t) + k_n(t) & -k_n(t) \\ & & & -k_n(t) & k_n(t) \end{bmatrix} \tag{5.3}
$$

In general, $R(t)$ can be any kind of loading; however, in this study, we assume that it is a harmonic force applied to the top floor:

$$
R(t) = \begin{bmatrix} 0 \\ \vdots \\ 0 \\ a \sin \omega t \end{bmatrix} \tag{5.4}
$$

where a and ω are the amplitude and frequency of excitation, respectively. For the sake of simplicity, in this study, we neglect damping effects.

To numerically solve the set of ordinary differential equations, Newmark explicit integrator is employed. To write the equations in the discrete state-space form, we introduce an extended state, z, that at each time instant t_k includes u, $\dot{u}$ and $\ddot{u}$ according to:

$$
z_k = \begin{bmatrix} u_k \\ \dot{u}_k \\ \ddot{u}_k \end{bmatrix}. \tag{5.5}
$$

Then state-space form of Eq. (5.1) is written as:

$$
z_k = A_k z_{k-1} + B_k \tag{5.6}
$$

where:

$$
A_k = \begin{bmatrix} I - \beta \Delta t^2 K_k M^{-1} & \Delta t I - \beta \Delta t^2 M^{-1}(D + \Delta t K_k) & -\beta \Delta t^2 M^{-1}\left(\Delta t^2\left(\tfrac{1}{2} - \beta\right)K_k + \Delta t(1 - \gamma)D\right) + \Delta t^2\left(\tfrac{1}{2} - \beta\right)I \\ -\gamma \Delta t K_k M^{-1} & I - \gamma \Delta t M^{-1}(D + \Delta t K_k) & -\gamma \Delta t M^{-1}\left(\Delta t^2\left(\tfrac{1}{2} - \beta\right)K_k + \Delta t(1 - \gamma)D\right) + \Delta t(1 - \gamma)I \\ -K_k M^{-1} & -M^{-1}(D + \Delta t K_k) & -M^{-1}\left(\Delta t^2\left(\tfrac{1}{2} - \beta\right)K_k + \Delta t(1 - \gamma)D\right) \end{bmatrix} \tag{5.7}
$$

and:

$$
B_k = \begin{bmatrix} \beta \Delta t^2 M^{-1} R_k \\ \gamma \Delta t M^{-1} R_k \\ M^{-1} R_k \end{bmatrix} \tag{5.8}
$$

β and γ are parameters of the Newmark algorithm, for details see Sect. 2.6.

Concerning the observation process, it is assumed that a part of state vector is directly observable; hence, observation equation is expressed as:

$$
y_k = H z_k + w_k \tag{5.9}
$$

where H denotes a Boolean matrix of appropriate dimension which links the states of the system to observation process, and w_k denotes associated measurement noise.

5.3 Reduced Order Modeling of Structural Systems

A detailed study of the application of POD for model order reduction of structural system has been presented in Chap. 3. However, to keep this chapter self-contained, in this Section, we review key features of the procedure. Let us assume that the displacement field $u \in \mathbb{R}^m$ of the system can be written in a separable form, according to:

$$u(x,t) = \sum_{i=1}^{m} \varphi_i(x)\alpha_i(t) \qquad (5.10)$$

where $\varphi_i(x)$ are a set of orthonormal vectors that satisfy proper orthogonal decomposition (POD) requirements and, α_i are temporal functions. Dealing with structural problems with high dimensional state vectors, the main variation in the data is usually occurring in a rather small subspace; consequently, it is frequently possible to approximate the state of the system by keeping just a few, say l proper orthogonal modes, with $l \ll m$:

$$u(x,t) \approx \sum_{i=1}^{m} \varphi_i(x)\alpha_i(t)$$
$$= \Phi_l \alpha \qquad (5.11)$$

where Φ_l denotes the matrix containing the retained l POMs of the system.

Substituting (5.11) into (5.1), and applying Galerkin projection yield the reduced dynamic model of the system:

$$M_l\ddot{\alpha} + D_l\dot{\alpha} + K_l\alpha = R_l(t) \qquad (5.12)$$

where:

$$M_l = \Phi_l^T M \Phi_l, \; D_l = \Phi_l^T D \Phi_l, \; K_l = \Phi_l^T K \Phi_l, \; R_l(t) = \Phi_l^T R(t). \qquad (5.13)$$

The reduced dynamic model in state-space form then reads:

$$z_{r,k} = A_k z_{r,k} + B_k + v_k^z \qquad (5.14)$$

$$y_k = HC z_{r,k} + w_k \qquad (5.15)$$

where the reduced order state includes the temporal coefficient, its first and second time derivatives:

$$z_{r,k} = \begin{bmatrix} \alpha_k \\ \dot{\alpha}_k \\ \ddot{\alpha}_k \end{bmatrix}. \tag{5.16}$$

In (5.14):

$$A_k = \begin{bmatrix} I - \beta \Delta t^2 M_l^{-1} K_{l,k} & \Delta t I - \beta \Delta t^2 M_l^{-1}(D_l + \Delta t K_{l,k}) & -\beta \Delta t^2 M_l^{-1}\left(\Delta t^2\left(\frac{1}{2} - \beta\right)K_{l,k} + \Delta t(1 - \gamma)D_l\right) + \Delta t^2\left(\frac{1}{2} - \beta\right)I \\ -\gamma \Delta t M_l^{-1} K_{l,k} & I - \gamma \Delta t M_l^{-1}(D_l + \Delta t K_{l,k}) & -\gamma \Delta t M_l^{-1}\left(\Delta t^2\left(\frac{1}{2} - \beta\right)K_{l,k} + \Delta t(1 - \gamma)D_l\right) + \Delta t(1 - \gamma)I \\ -M_l^{-1} K_{l,k} & -M_l^{-1}(D_l + \Delta t K_{l,k}) & -M_l^{-1}\left(\Delta t^2\left(\frac{1}{2} - \beta\right)K_{l,k} + \Delta t(1 - \gamma)D_l\right) \end{bmatrix} \tag{5.17}$$

$$B_{l,k} = \begin{bmatrix} \beta \Delta t^2 M_l^{-1} R_{l,k} \\ \gamma \Delta t M_l^{-1} R_{l,k} \\ M_l^{-1} R_{l,k} \end{bmatrix} \tag{5.18}$$

and, in (5.15):

$$C = \begin{bmatrix} \Phi_l & & \\ & \Phi_l & \\ & & \Phi_l \end{bmatrix} \tag{5.19}$$

Throughout the paper, whenever two indexes are used to denote a variable, the first subscript (r) refers to a property associated with reduced order model, while the second subscript refers to the time instant at which variable is considered.

In (5.14) and (5.15), v_k^z and w_k are the process and measurement noises, respectively. The former uncertainty stems from the loss of accuracy due to the reduced modeling which needs to be further assessed in order to determine its probability distribution and verify the correlation structure in it. In Chap. 4, we have tested the whiteness of the residual error signal of POD-based reduced model of Pirelli tower; it has been shown that, by an increase in the number of POMs retained in the analysis, a reduction occurs in the amplitude of the noise signal and its spectral power. As a consequence, the effect of the non-white uncertainty in the Kalman-POD observer becomes negligible. Hence in this chapter, we assume that the noises satisfy the requirements of the family of recursive Bayesian inference algorithms.

To tackle the dual estimation problem, we now augment the parameters of the reduced model into the state vector, to comply with the state space form. Subsequently, we introduce the augmented state vector $x_{r,k}$, that at any time t_k encompasses both states and parameters of the system $x_{r,k} = [z_{r,k} \quad \vartheta_{r,k}]^T$. In Sect. 2.2, it is shown that dual estimation of states and parameters of a linear system leads to a nonlinear state-space model. The new state space equation is written as:

$$x_{r,k} = f_{r,k}(x_{r,k-1}) + v_k \tag{5.20}$$

$$y_k = H L x_{r,k} + w_k \tag{5.21}$$

$$L = \begin{bmatrix} C \\ & 0 \end{bmatrix} \qquad (5.22)$$

where: 0 in L is a null matrix of appropriate dimension to annihilate the effects in the observation mapping of parameters in the augmented state vector; $f_{r,k}(.)$ maps the state of the system in time and H denotes the correlation between states and observables of the system; L links the reduced states of the system to the full state; whereas v_k and w_k stand for the zero mean white Gaussian processes with associated covariance matrices V and W. Likewise previous Chapters, $\vartheta_{r,k}$ includes the parameters of the reduced state space model that should be estimated, namely the components of the reduced stiffness matrix $K_{l,k}$.

5.4 Dual Estimation of Reduced States and Parameters of a Damaging Structure

Dual estimation problem for a non-damaging (elastic) structure can be pursued via the estimation of reduced state and parameters since there will not be changes in the subspace of the problem. On the contrary, subspace of a damaging structure varies in time: for instance, a change in a story stiffness can lead to a change in the POMs. As a consequence, dual estimation of the reduced state and parameters of a damaging structure not only includes tracking of the reduced state and estimation of the reduced parameters of the system, but also needs online update of the relevant subspace of the structure.

In this section, we introduce a novel approach for simultaneous state and parameter estimation, accompanied by an online subspace update in order to obtain an estimate of the full state. In this regard, we adopt recursive Bayesian filters: the extended Kalman filter (EKF) and the extended Kalman particle filter (EK-PF). They have been discussed in Chap. 2, and used for dual estimation. A Kalman filter is instead used to update the subspace furnished by POD. Likewise all recursive Bayesian inference algorithms, the iterations start by an initial guess; next, within each time interval $[t_{k-1} t_k]$, provided that at t_{k-1} estimations of state, parameters and subspace of the system are available, the state $z_{r,k}$ and parameters in $K_{l,k}$ are simultaneously estimated. Let us consider the following state space model:

$$x_{r,k} = f_{r,k}(x_{r,k-1}) + v_k \qquad (5.23)$$

$$y_k = HL_k x_{r,k} + w_k \qquad (5.24)$$

where:

$$L_k = \begin{bmatrix} \Phi_{l,k} \\ & \Phi_{l,k} \\ & & \Phi_{l,k} \\ & & & 0 \end{bmatrix}. \qquad (5.25)$$

Along with Eqs. (5.23) and (5.24), an additional equation should be introduced in order to permit time variation and update of $\boldsymbol{\Phi}_l$, similar to the trick used for dual estimation of states and parameters. The following equation is introduced to allow the subspace to vary over time, and use the data in observation in order to adapt to the possible changes:

$$\boldsymbol{\Phi}_{l,k} = \boldsymbol{\Phi}_{l,k-1} + \upsilon \qquad (5.26)$$

where υ denotes a fictitious zero mean, white Gaussian noise with associated covariance υ that needs to be obviously tuned to obtain unbiased estimates of the subspace vectors.

To recursively update the subspace, Eqs. (5.26) and (5.24) are assumed as the state-space model for subspace evolution. The former equation governs the evolution of the subspace, and the latter one links the observation to the subspace. In Eqs. (5.26) and (5.24), it is assumed that $\boldsymbol{x}_{r,k}$ remains independent of $\boldsymbol{\Phi}_{l,k}$. The observation Eq. (5.24), when used for subspace update can be rewritten as:

$$\boldsymbol{y}_k = \boldsymbol{H}_{ss}\boldsymbol{\Phi}_{l,k} + \boldsymbol{w}_k \qquad (5.27)$$

where $\boldsymbol{H}_{ss}$ is a stationary matrix which links the observation process to the subspace spanned by the POMs, and can be computed by manipulating Eq. (5.26). Equation (5.27) establishes a linear relationship between the observation $\boldsymbol{y}_k$ and the subspace $\boldsymbol{\Phi}_{l,k}$, whose linearity allows us to use the Kalman filter (the optimal estimator for linear state-space models) for the estimation of the subspace.

In Tables 5.1 and 5.2, an algorithmic description of the procedure is reported; the EKF and the EK-PF are used for dual estimation. In the Table 5.1, $\nabla_x \boldsymbol{f}_{r,k}(x)|_{x=\widehat{\boldsymbol{x}}_{k-1}}$ denotes Jacobian of $\boldsymbol{f}_{r,k}(\blacksquare)$, at $\boldsymbol{x}_r = \boldsymbol{x}_{r,k}^-$.

As seen in Table 5.1, the algorithm has two main stages of prediction and update. In the prediction stage, the evolution equations are used to map in time the reduced state $\boldsymbol{x}_{r,k-1}$ along with its covariance. In the update stage, first the reduced state and parameters and their associated covariances are corrected by incorporating the information contained in the latest observation (steps 1 and 2); next, the Kalman filter is exploited to update the subspace $\boldsymbol{\Phi}_l$. Step 3 in the prediction stage of dual estimation algorithm is in fact the predictor stage of the Kalman filter to update the subspace. In step 4, Kalman gain is computed and is used in step 5 to update the estimate of the subspace by taking the latest observation into account.

Concerning the use of EK-PF for dual estimation, according to previous Chap. 2, combined with the Kalman filter for subspace update, similar to the procedure used by EKF-KF algorithm, the reader is referred to Table 5.2. In the Table 5.2, $\boldsymbol{F}_{r,k}^{(i)}$ is:

$$\nabla_x \boldsymbol{f}_{r,k}(\boldsymbol{x})|_{\boldsymbol{x}=\widehat{\boldsymbol{x}}_{k-1}} \qquad (5.28)$$

where it denotes Jacobian of the reduced evolution $\boldsymbol{f}_{r,k}(\boldsymbol{x}_r)$ at $\boldsymbol{x}_r = \boldsymbol{x}_{r,k}^{(i)-}$.

Table 5.1 EKF-KF algorithm for dual estimation of the reduced model and subspace update

---

- Initialization at time t_0

$$\hat{x}_{r,0} = L_0^T \mathbb{E}[x_0] \qquad P_{r,0} = L_0^T \mathbb{E}\left[(x_0 - \hat{x}_0)(x_0 - \hat{x}_0)^T\right] L_0$$

$$\hat{\Phi}_{l,0} = \mathbb{E}[\Phi_{l,0}] \qquad P_{ss,0} = \mathbb{E}\left[\left(\Phi_{l,0} - \hat{\Phi}_{l,0}\right)\left(\Phi_{l,0} - \hat{\Phi}_{l,0}\right)^T\right]$$

- At time t_k, for $k = 1, \ldots, N_t$
  - Prediction stage
    1. Computing process model Jacobian

    $$F_{r,k} = \nabla_x f_{r,k}(x)\big|_{x = \hat{x}_{k-1}}$$

    2. Evolution of state and prediction of covariance

    $$x_{r,k}^- = f_{r,k}(x_{r,k-1})$$
    $$P_{r,k}^- = F_{r,k} P_{r,k-1}^- F_{r,k}^T + V$$

  - Update stage
    1. Use $\Phi_{l,k-1}$ to estimated L_k and Kalman gain

    $$G_k = P_{r,k}^- L_k^T H^T \left(H L_k P_{r,k}^- L_k^T H^T + W\right)^{-1}$$

    2. Update state and covariance

    $$x_{r,k} = x_{r,k}^- + G_k\left(y_k - H L_k x_{r,k}^-\right)$$
    $$P_{r,k} = P_{r,k}^- - G_k H L_k P_{r,k}^-$$

    3. Predict subspace and its associated covariance

    $$\Phi_{l,k}^- = \Phi_{l,k-1}$$
    $$P_{ss,k}^- = P_{ss,k-1} + \Upsilon$$

    4. Calculate Kalman gain for updating subspace

    $$G_{ss,k} = P_{ss,k}^- H_{ss}^T \left(H_{ss} P_{ss,k}^- H_{ss}^T + W\right)^{-1}$$

    5. Calculate Kalman gain for updating subspace

    $$\Phi_{l,k} = \Phi_{l,k}^- + G_{ss,k}\left(y_k - H_{ss}\Phi_{l,k}^-\right)$$
    $$P_{ss,k} = P_{ss,k}^- - G_{ss,k} H_{ss} P_{ss,k}^-$$

5.5 Numerical Results: Damage Detection in a Ten Storey Shear Building

This section deals with the numerical assessment of the proposed algorithm to detect damage in a 10-storey shear building. To deal with the damage scenarios, it is not straight forward to use the model of Pirelli tower, due to the fact that a static condensation has been carried out to derive matrices of lumped mass system of the Pirelli towers. For the sake of simplicity, in the numerical example, it is assumed that all the floors have equal mass and inter-storey stiffness, i.e. $m_i = 20$ Kg and $k_i = 300$ Kg/m where $i = 1, 2, \ldots, 10$, and the damping effect is neglected. It the analysis, the external load shaking the structure, is a sinusoidal load applied to the last floor (roof) of the building, varying according to:

$$R(t) = a_m \sin 2\pi\omega t \tag{5.29}$$

Table 5.2 EK-PF-KF algorithm for dual estimation of the reduced model and subspace update

- Initialization at time t_0

$$\widehat{x}_{r,0} = L_0^{\mathrm{T}} \mathbb{E}[x_0] \qquad P_{r,0} = L_0^{\mathrm{T}} \mathbb{E}\left[(x_0 - \widehat{x}_0)(x_0 - \widehat{x}_0)^{\mathrm{T}}\right] L_0$$

$$\widehat{\Phi}_{l,0} = \mathbb{E}[\Phi_{l,0}] \qquad P_{ss,0} = \mathbb{E}\left[\left(\Phi_{l,0} - \widehat{\Phi}_{l,0}\right)\left(\Phi_{l,0} - \widehat{\Phi}_{l,0}\right)^{\mathrm{T}}\right]$$

$$x_{r,0}^{(i)} = \widehat{x}_0 \qquad \omega_0^{(i)} = p(y_0|x_{r,0}), \ i = 1, \dots, N_{\mathrm{P}}$$

- At time t_k, for $k = 1, \dots, N_t$
- Prediction stage
  1. Draw particles

$$x_{r,k}^{(i)-} \sim p\left(x_{r,k}|x_{r,k-1}^{(i)}\right) \quad i = 1, \dots, N_{\mathrm{P}}$$

  2. Push the particles toward the region of high probability through an EKF

$$P_{r,k}^{(i)-} = F_{r,k}^{(i)} P_{r,k-1}^{(i)} F_{r,k}^{(i)T} + V$$

$$G_k^{(i)} = P_{r,k}^{(i)-} L_{k-1}^{\mathrm{T}} H_k^{\mathrm{T}} \left(HL_{k-1} P_{r,k}^{(i)-} L_{k-1}^{\mathrm{T}} H^{\mathrm{T}} + W\right)^{-1}$$

$$x_{r,k}^{(i)} = x_{r,k}^{(i)-} + G_k^{(i)}\left(y_k - HL_{k-1} x_{r,k}^{(i)-}\right)$$
$$P_{r,k}^{(i)} = P_{r,k}^{(i)-} - G_k^{(i)} HL_{k-1} P_{r,k}^{(i)-} \qquad i = 1, \dots, N_{\mathrm{P}}$$

- Update stage
  1. Evolve weights

$$\omega_k^{(i)} = \omega_{k-1}^{(i)} p\left(y_k|x_{r,k}^{(i)}\right) \quad i = 1, \dots, N_{\mathrm{P}}$$

  2. Resampling, see Table 2.5.
  3. Compute expected value or other required statistics

$$\widehat{x}_{r,k} = \sum_{i=1}^{N_{\mathrm{P}}} \omega_k^{(i)} x_{r,k}^{(i)}$$

  4. Predict subspace and its associated covariance

$$\Phi_{l,k}^- = \Phi_{l,k-1}$$
$$P_{ss,k}^- = P_{ss,k-1} + \Upsilon$$

  5. Calculate Kalman gain for updating subspace

$$G_{ss,k} = P_{ss,k}^- H_{ss}^{\mathrm{T}} \left(H_{ss} P_{ss,k}^- H_{ss}^{\mathrm{T}} + W\right)^{-1}$$

  6. Update subspace and its associated covariance

$$\Phi_{l,k} = \Phi_{l,k}^- + G_{ss,k}\left(y_k - H_{ss}\Phi_{l,k}^-\right)$$
$$P_{ss,k} = P_{ss,k}^- - G_{ss,k} H_{ss} P_{ss,k}^-$$

where $a_m = 10\,\mathrm{N}$ and $\omega = 0.01\,\mathrm{Hz}$.

Consider a case in which a stiffness reduction equal to 50 % has occurred at the 5th floor. The POMs of the structure, before and after damage occurrence, are computed and presented in the Fig. 5.2. To compute these POMs of the healthy and damaged cases, two direct analyses have been carried out to assemble the so-called snapshot matrices. Looking at Fig. 5.2, it can be seen that the ten POMs of the structure are affected by the stiffness reduction at the 5th floor. The effect of the

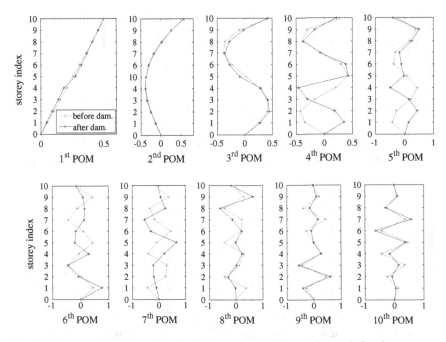

Fig. 5.2 Proper orthogonal modes of a 10 storey shear building before and after damage

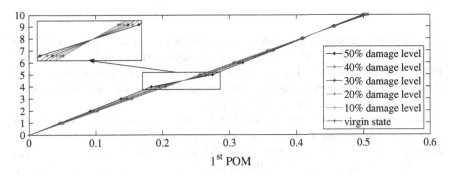

Fig. 5.3 1st POM of the 10 storey shear building subject to different levels of damage at 5th floor

damage in the first POM is quite visible, the usefulness of such sensitivity to damage, even in the first POM, helps tracking the evolution of damage in a single DOF reduced model.

Figure 5.3 compares the first POM of the structure when the 5th floor of the structure suffers a damage of varying intensity; the close-up in the graph allows us to compare the shape of the POM in the vicinity of the damage location. Obviously, the intensity of damage leads to an increase in the deviation of the POM

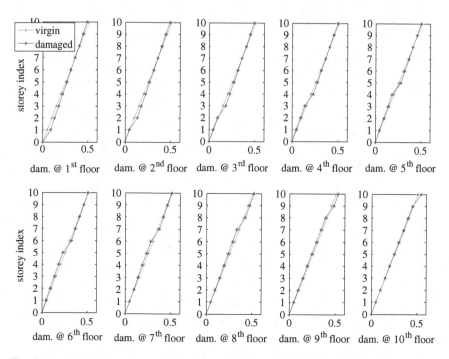

Fig. 5.4 1st POM of a ten storey shear building for a damage occurring at different storeys of the building

relevant to the damaged state with respect to the healthy state of the structure. To highlight the sensitivity of the 1st POM to damage location, in Fig. 5.4 the first POM of the damaged state is compared with healthy state, when damage occurs at different floors. The imposed level of the damage in all the cases is equal to a 50 % reduction of the stiffness of the relevant floor.

Now that the link between the first POM of the structure and the location and severity of the damage is established, we move to the problem of the recursive estimation of the state, parameters and POMs of the reduced model of the structure. To detect the damage, the POMs of healthy and current state of the structure are compared; thus information concerning the healthy state of the structure is needed. In this study, the case in which the reduced models retain one or two POMs are assessed, the latter case is mainly reported to verify the performance of the algorithm in case of the higher number of parameters to be estimated: dual estimation of reduced models which retain more POMs includes calibrating a high number of parameters, and can therefore potentially pose the problem of curse of dimensionality, as discussed in Chap. 2.

First, we deal with the reduced model constructed through a single POM. Pseudo-experimental data for evaluation of the methodology have been created by running direct analysis, to compute the response of the structure, and then adding zero mean white Gaussian noise to allow for uncertainties in measuring the

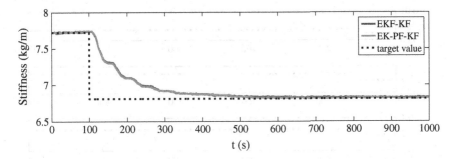

Fig. 5.5 Estimation of the reduced via EKF-KF and EK-PF-KF algorithms

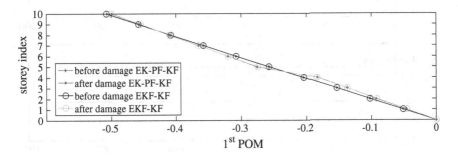

Fig. 5.6 1st POM of the structure estimated EKF-KF and EK-PF-KF algorithms

response of the structure. The covariance of the added noise to all the pseudo experimental data considered in this section is set to 10^{-4} m^2 to simulate a high level of measurement uncertainty. The duration of the analysis is set to 1,000 s, in order to allow the estimates of the algorithms to converge to a steady state value. The damage scenario is once again a reduction of 50 % in the stiffness of the 5th floor, which occurs at $t = 100$ s. Other damage scenarios, featuring severities ranging from 10 to 40 % in the reduction of the stiffness of other floors has been assessed; the algorithms show similar performance dealing with those scenarios; thus the results are not presented for the sake of brevity.

Since the goal of this Section is the damage identification, the results concerning the estimation of the state are not discussed. Figure 5.5 shows the time history of the estimated stiffness of the reduced system when compared with its target value. It is observed that before damage occurs, the estimation coincides with the target value; however, after damage occurs, it takes almost 400 s for the algorithm to make its estimate to converge to the target value. Figure 5.6 shows the estimated POMs of the building before and after damage: the POM concerning the healthy state is related to $t = 50$ s, and the POM concerning the damaged state is related to $t = 1,000$ s. To compare the performance of the algorithm in tracking the POM of the system over time, Fig. 5.7 shows time history of the estimated POM, compared with its target value. It is observed that the estimations of the

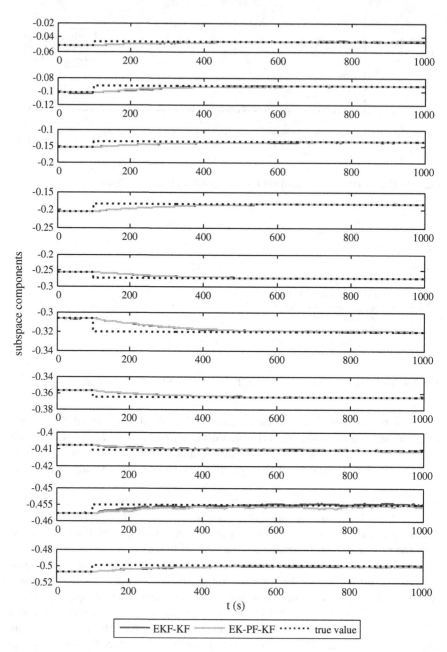

Fig. 5.7 Time histories of the components of the 1st POM of the structure, from *top* to *bottom* respectively corresponds to first to last component of the POM vector (time histories of entries of POM)

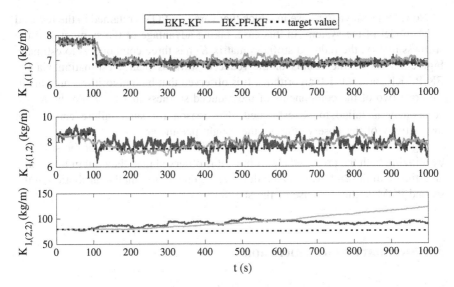

Fig. 5.8 Time histories of the parameter estimation of the reduced model via EK-PF-KF and EKF-KF algorithms: $K_{l,(1,1)}$, $K_{l,(1,2)}$ and $K_{l,(2,2)}$ from *top* to *bottom*, respectively

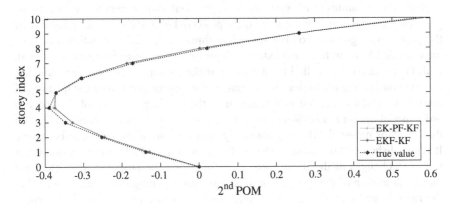

Fig. 5.9 Results concerning estimation of the second POM of the shear building after damage occurs

POM components before damage occurrence coincide with the true value; after damage occurs, the algorithm needs nearly 400 s, similar to parameter estimates, to reach steady state. EK-PF, when dealing with several problems discussed in Chap. 2 outperforms the EKF; hence, it is used to verify if its convergence rate would be better than EKF's one. However, it is seen in Fig. 5.5 that the quality of estimation of the reduced stiffness and the 1st POM of the structure do not change, when either EKF-KF or EK-PF-KF are used for dual estimation and reduced order modeling of the damaging shear building.

Now, let us move to a case in which there are two POMs retained in the reduced order model of the system. In this case, taking advantage of the symmetry of the stiffness matrix, the reduced stiffness matrix K_l has three components to estimate. Figure 5.8 shows the results of the reduced stiffness matrix estimation via the EK-PF-KF and EKF-KF algorithms. It is observed that both algorithms are able to calibrate two of the components of the reduced stiffness matrix, while the $K_{l,(2,2)}$ component is failed to be estimated. The reason for such failure can be the insensitivity of the observations to the sought parameter.

Figure 5.9 shows the results of the estimation of the 2nd POM of the structure by utilizing both the proposed algorithms. It is observed that, they fail in furnishing an estimate of the 2nd POM; this failure can be due to the small contribution of the second POM in the response of the structure.

5.6 Summary and Conclusion

In this chapter, we consider dual estimation and reduced order modeling of a damaging structure. Moreover, proper orthogonal decomposition has been considered for reduced order modeling in order to find a subspace which optimally captures the dynamics of the system. Through a Galerkin projection, the equations governing the dynamics of the system are projected onto the subspace provided by the proper orthogonal decomposition algorithm. As for the dual estimation goal, the extended Kalman filter and extended Kalman particle filter have been adopted; both filters, in their so-called update stage, make a comparison between the latest observation and the prediction of the state of the system to estimate the quantity of correction which is needed in estimation of the state. In the case of the reduced order modeling, for realization of such a comparison, reconstruction of full state of the system is required, which is obviously possible only if the subspace is known. It is established that the subspace found by proper orthogonal decomposition is not robust to changes of the parameters; therefore, we have proposed algorithms for online estimation of the subspace spanned by proper orthogonal modes retained in the reduced order model of the system. Such an online estimation of the proper orthogonal modes of the structure makes it possible to detect the damage in the structure, locate it and potentially identify its intensity.

References

Abazarsa F, Ghahari SF, Nateghi F, Taciroglu E (2013a) Response-only modal identification of structures using limited sensors. Struct Control Health Monit 20:987–1006

Abazarsa F, Nateghi F, Ghahari SF, Taciroglu E (2013b) Blind modal identification of non-classically damped systems from free or ambient vibration records. Earthq Spectra. doi:http://dx.doi.org/10.1193/031712EQS093M (in press)

Galvanetto U, Violaris G (2007) Numerical investigation of a new damage detection method based on proper orthogonal decomposition. Mech Syst Signal Process 21:1346–1361

Ghahari SF, Abazarsa F, Ghannad MA, Celebi M, Taciroglu E (2013a) Blind modal identification of structures from spatially sparse seismic response signals. Struct Control Health Monit. doi:10.1002/stc.1593 (in press)

Ghahari SF, Abazarsa F, Ghannad MA, Taciroglu E (2013b) Response-only modal identification of structures using strong motion data. Earthquake Eng Struct Dynam 42:1221–1242

Ghahari SF, Ghannad MA, Taciroglu E (2013c) Blind identification of soil-structure system. Soil Dynamics Earthq Eng 45:56–69

Ghahari SF, Ghannad MA, Norman J, Crewe A, Abazarsa F, Taciroglu E (2013d) Considering wave passage effects in blind identification of long-span bridges. Topics in model validation and uncertainty quantification, vol 5. Springer, New York, pp 53–66

Loh C, Weng J, Liu Y, Lin P, Huang S (2011) Structural damage diagnosis based on on-line recursive stochastic subspace identification. Smart Mater Struct 20:055004

Moaveni B, He X, Conte JP, Restrepo JI, Panagiotou M (2011) System identification study of a 7-story full-scale building slice tested on the UCSD-NEES shake table. J Struct Eng 137:705–717

Ruotolo R, Surace C (1999) Using SVD to detect damage in structures with different operational conditions. J Sound Vib 226:425–439

Shane C, Jha R (2011) Proper orthogonal decomposition based algorithm for detecting damage location and severity in composite beams. Mech Syst Signal Process 25:1062–1072

Van Overschee P, De Moor B (1996) Subspace identification for linear systems. Kluwer Academic, Dordrecht

Vanlanduit S, Parloo E, Cauberghe B, Guillaume P, Verboven P (2005) A robust singular value decomposition for damage detection under changing operating conditions and structural uncertainties. J Sound Vib 284:1033–1050

Yang JN, Lin S (2005) Identification of parametric variations of structures based on least squares estimation and adaptive tracking technique. J Eng Mech 131:290–298

Chapter 6
Conclusions

Abstract The major goal of the present study is to develop fast and robust algorithms for online damage detection in structural systems. To accomplish this objective, the research study presented in this monograph can contribute to three different research areas: (a) stochastic system identification of multi degrees-of-freedom structural systems via recursive Bayesian inference algorithms, (b) reduced order modeling of multi degrees-of-freedom structural systems through proper orthogonal decomposition; and (c) stochastic system identification of reduced order models of multi degrees of freedom structural systems through recursive Bayesian filters.

6.1 Summary of Contributions

The principal contributions and major findings of this research study can be summarized as follows:

(1) Four state of the art Bayesian filters, namely the extended Kalman filter, the sigma-point Kalman filter, the particle filter and the hybrid extended Kalman particle filter have been adopted. To benchmark the performance of filters and avoid shadowing effects of the structure, the filters have been adopted to recursively identify the parameters of the constitutive model of a single degree-of-freedom dynamical system: an exponential softening, and three bilinear models (linear-hardening, linear-plastic and linear-softening), as possible representatives of initial stages of damage are adopted. The goal is achieved by dual estimation concept, where the parameters of the system are joined to the state vector in order to simultaneously track the state of the system and calibrate the parameters, as new observations become available. Provided that the Jacobian of the evolution equations of the state space model is positive definite and bounded, it is known that the adopted filters are stable and can converge to unbiased estimates; however, such conditions are not

S. Eftekhar Azam, *Online Damage Detection in Structural Systems*,
PoliMI SpringerBriefs, DOI: 10.1007/978-3-319-02559-9_6,
© The Author(s) 2014

always satisfied in a model featuring softening constitutive law. This fact substantiates numerical assessment of stability and convergence of the studied filters, when applied to the estimation of parameters of a softening constitutive law used to describe damage evolution in the system. The conducted numerical campaign has revealed that while the extended Kalman filter, the unscented Kalman filter and the particle filter all fail to provide unbiased estimates of the sought parameters, the hybrid extended Kalman particle filter performs rather reasonably.

(2) The extended Kalman filter (because of its computational time efficiency) and the hybrid extended Kalman particle filter (due to its excellent performance when applied to the analysis of single degree-of-freedom nonlinear system) have been adopted for dual estimation of states and constitutive parameters of a multi degrees-of-freedom linear shear building-type structure. The performance of the two filters has been assessed through the estimation of the values of the inter-storey stiffness of the floors of the building. In the simplest case, i.e. a two-storey shear building, both filters furnish quite accurate estimates of the stiffness values; however, moving to a three-storey structure, the performance of both filters is adversely affected. The trend is corroborated by the results in the case of a four storey building: the estimation resulted in a bias up to 50 % of the target values of the parameters. This trend suggests that, when dealing with dual estimation of a multi storey shear building, an increase in the number of storeys rapidly deteriorates accuracy of the parameter estimates. Therefore, this approach will not be an effective damage detection method; thus we the adopted a dual estimation of a reduced order model of the building.

(3) To manage the curse of dimensionality issue, the method of proper orthogonal decomposition (POD) has been adopted to produce a reduced order model of the vibrating structure. Provided that there exist a set of samples from the response of the system and its members are selected in way that the ensemble contains information on the main dynamic characteristics of the system; thus POD automatically looks for those main characteristics. To accomplish this objective, the POD finds the directions which capture the maximum variation, or equivalently, the maximum energy of the system. Once the relevant directions (called proper orthogonal modes, POMs) in an initial training stage are found, Galerkin projection is employed to project the equations onto the subspace spanned by the computed POMs. The efficiency of the algorithm in terms of speed-up and accuracy of the estimations has been then numerically assessed. The procedure is applied for reduced order modeling of the Pirelli tower located in Milan; prediction capability and speed-up issues are numerically assessed. It is observed that reducing the original 39 degrees-of-freedom structure to a reduced model consisting of four POMs makes the computations 250 times faster; while a reduced model featuring a single POM has a speed-up value of 500. Moreover, robustness of the reduced models, featuring different number of retained POMs, to a change in the source of the external loading has been further analyzed. To produce the samples required

for initial training stage of POD-based reduced model, the Pirelli tower has been assumed to be shaken by the well-known El Centro acceleration time history. The resulted reduced model has then been applied to simulate the response of the structure to the Kobe and Friuli earthquake excitations. It has been shown that the change in the source of excitation does not affect much the prediction capabilities of POD-based reduced models in seismic analysis of the structure.

(4) Prior to applying the reduced models obtained by the POD in the recursive Bayesian inference algorithms adopted in this monograph, a statistical assessment of the uncertainties induced by reduced order modeling is essential. In this study, all the Bayesian filters adopted are assumed that the uncertainties in the state space model are uncorrelated processes. The null hypothesis of whiteness of the residual error of POD models has been tested by cumulative periodogram-based test of Bartlett (Bartlett 1978). It has been shown that, no matter what the number of the POMs featured by the reduced model is, its residual error is always correlated. However, by an increase in the number of retained POMs, the spectral power of the correlation in the signal decreases. The linear, time-invariant reduced models of the Pirelli tower has been incorporated into a Kalman filter in order to speed-up the calculations. Provided that the noises in the state space equation are white Gaussian processes, it is known that Kalman filter furnishes optimal estimates of state of a linear model. We have shown that the POD-based reduced state space used in this study is not white. That is, when just a single POM is retained in the analysis, residual mean squared error (RMSE) of the POD-Kalman observer is higher than the POD alone; however, as the number of POMs retained in the analysis increases and spectral power of the correlations decrease, POD-Kalman observer performs better, in terms of reducing RMSE of estimates: POD-Kalman observer featuring three and four POMs in its reduced model decrease quality of estimates provided by POD alone. Concerning speed-up gained by introducing POD-based models into Kalman observer, by maintaining a minimal number of POMs, the observer is run up hundreds of times faster.

(5) Besides its efficiency in model order reduction, the POD has an interesting feature which makes it appropriate for the purpose of damage detection. Proper orthogonal modes which are furnished by the POD have been shown to be sensitive to the severity and location of the damage in the mechanical systems, and they are already used as damage detection tools (Shane and Jha 2011a). These two aspects of POD, namely its efficiency for model order reduction and its capability in identifying the damage, make an ideal candidate for the problem of damage detection in structural systems via reduced order modeling and dual estimation. In this monograph, we have proposed a novel algorithm for dual estimation of a POD-based reduced order model of a time-varying shear model of building. The capability of the algorithm in tracking the state of the system, the parameters of the reduced model and the POMs of the reduced model has been numerically assessed. Our approach has been

employed to detect a variety of damage scenarios in a ten-storey shear building; however, the assessment has been based on pseudo experimental verifications. It has been concluded that the proposed procedure performs accurately.

The major goal of this monograph is to develop robust algorithms for online and real-time detection of the damage in civil structures. The objective of the monograph is perceived by developing a procedure by a synergy of recursive Bayesian inference methods and proper orthogonal decomposition. Therefore, a POD-based reduced model of the structure has been considered: dual estimation concept has been exploited, within a recursive Bayesian framework the state and the parameters of the reduced model are simultaneously estimate based on observational signal which becomes available in discrete time instants. In each recursion, not only the state and the parameters of the reduced model are estimated, but also the proper orthogonal modes employed to construct the reduced model are estimated. It is shown that the POD modes can indicate location and severity of the damage in mechanical systems. The unbiased estimate of the POMs provided by our approach permits robust, online and real-time indication of the damage in a shear type of building.

6.2 Suggestions for Future Research

Based on the work presented herein, several research areas have been identified as open to and in need of future work:

(1) In this monograph, regarding the application of Bayesian filters for dual estimation of states and parameters of the multi-storey shear buildings, we have adopted the family of Kalman filter, particle filter and a combination thereof. However, the use of evolutionary particle filters has not been considered; it is suggested to tackle this problem by utilizing the aforementioned filters as well.

(2) To construct the POD-based reduced models, the effects of nonlinear mechanisms have been neglected. It is recommended to take those effects into consideration as well.

(3) The algorithms proposed in this monograph for damage detection via dual estimation of the reduced model and subspace update have been assumed to be fed by displacement response at each floor. The reason is to construct the reduced model POD modes of the displacement response of the structure used for acceleration modes are different from displacement modes; moreover, the accuracy of reproducing accelerations by reduced model is lower than displacements. There are two remedies: one is increasing the number of POMs retained in the reduced model to improve the quality of acceleration reconstruction; hence, this can lead to curse of dimensionality by increasing number

of the parameters to be estimated in the reduced model, and the other option is to compute the displacement response from the acceleration response data. In the literature, there are several methods available to calculate displacement response based on the acceleration (Skolnik et al. 2011). It is recommended to utilize those techniques to verify the algorithms by pseudo experimental data. It is worthy to see if the Bayesian filters can handle the uncertainty introduced by converting the acceleration response into the floor displacements.

(4) Through this study, the methodologies which were used or developed have been verified via pseudo-experiments. It is recommended to verify the effectiveness of the proposed procedure by utilizing real experiments.

(5) It is has been shown that, dealing with a ten-storey shear building with equal masses and stiffnesses at each floor, there exist an intuitive and clear correlation between damage location and intensity and the POM. However, to quantify the damage index relevant to each floor, it is recommended to utilize artificial neural networks (the standard classification methodologies) in order to provide quantitative damage indexes for each storey based on the POM of the structure; such method has been already adopted to identify damage based on the changes in the coefficients of an auto regressive moving average model of a four storey structure (de Lautour and Omenzetter 2010).

References

Bartlett MS (1978) An introduction to stochastic processes with special reference to methods and applications. Cambridge University Press, London

de Lautour OR, Omenzetter P (2010) Damage classification and estimation in experimental structures using time series analysis and pattern recognition. Mech Syst Sign Proces 24:1556–1569

Shane C, Jha R (2011a) Proper orthogonal decomposition based algorithm for detecting damage location and severity in composite beams. Mech Syst Sign Proces 25:1062–1072

Skolnik DA, Nigbor RL, Wallace JW (2011) A quantitative basis for building instrumentation specifications. Earthquake Spectra 27:133–152

Appendix:
Summary of the Recursive Bayesian Inference Schemes

Abstract This appendix aims to provide a summary of the key aspects of general sequential Bayesian estimation problem and to highlight the reason for which a closed-form solution can be obtained when certain prerequisites are met. In addition, the reasons for which approximate solution are widely sought for are explained. In so doing, for the sake of the brevity the details are not presented and instead significant text and research books on the topic are cited (Bittanti 2004; Ljung 1999; Doucet et al. 2001; Haykin 2001), so that the reader can find the details on the derivation of the Bayesian algorithms used in this monograph.

While studying the dynamics of a structural system, usually we have to deal with a state space representation of it; let us consider the following state-space equation:

$$x_k = f_k(x_{k-1}) + v_k \tag{A.1}$$

$$y_k = H_k x_k + w_k \tag{A.2}$$

where $f_k(\blacksquare)$, maps the state vector x_k. over time, H_k. links the (usually unobservable or partially observable) state vector x_k to the observation y_k. The v_k and w_k denote the zero mean additive noises, which quantitatively represent the model and observation inaccuracies, respectively.

The inference problem can be viewed as recursively estimating the expected value $E[x_k|y_{1:k}]$ of the state of the system, conditioned on the observations. Provided that the initial probability density function (PDF) $p(x_0|y_0) = p(x_0)$ of the state vector is known, the problem consists in recursively estimating $p(x_k|y_{1:k})$ at the time t_k, assuming that the conditional probability density function $p(x_{k-1}|y_{1:k-1})$ is available at the time t_{k-1}. In the literature of the sequential Bayesian estimation, it is customary to estimate the state of a system in two different stages: prediction and update. In the prediction stage the well-known Chapman-Kolmogorov equation provides so called a-priory estimate of the PDF of the state at t_k (Arulampalam et al. 2002):

$$p(x_k|y_{1:k-1}) = \int p(x_k|x_{k-1}) p(x_{k-1}|y_{1:k-1}) dx_{k-1}. \tag{A.3}$$

In the update stage, by taking advantage of the Bayes rule, the PDF of the state is adjusted via including the information conveyed by the observation y_k (Cadini et al. 2009) in the estimation:

S. Eftekhar Azam, *Online Damage Detection in Structural Systems,*
PoliMI SpringerBriefs, DOI: 10.1007/978-3-319-02559-9,
© The Author(s) 2014

$$p(x_k|y_{1:k}) = \varsigma \, p(y_k|x_k)p(x_k|y_{1:k-1}) \tag{A.4}$$

where ς is a normalizing constant which depends on the likelihood function of the observation process. The Eqs. (A.3) and (A.4) together set the basis for any recursive Bayesian inference scheme. The analytical solution of the integral in the (A.3) is not possible, except for a limited category of systems, namely systems formulated by linear state-space equations and disturbed by white Gaussian noises (Julier and Uhlmann 1997). Provided that the probability density functions of the evolution and the observation equations are Gaussian, they both are represented by the following exponential $e^{(\blacksquare)}$ form:

$$p(x_{k-1}|y_{1:k-1}) = \frac{1}{((2\pi)^n|P_{k-1}|)^{1/2}} \exp\left[-\frac{1}{2}(x_{k-1} - \hat{x}_{k-1})^T P_{k-1}^{-1}(x_{k-1} - \hat{x}_{k-1})\right]$$

$$\tag{A.5}$$

$$p(x_k|x_{k-1}) = \frac{1}{((2\pi)^n|V|)^{1/2}} \exp\left[-\frac{1}{2}(x_{k-1} - f_k(x_{k-1}))^T V^{-1}(x_{k-1} - f_k(x_{k-1}))\right]$$

$$\tag{A.5}$$

where, P_{k-1} and V are the covariances of the estimated state and evolution uncertainty, respectively; n is the dimension of the stare vector. Provided that the evolution equation is linear, the integral in the A.3 can be dealt with analytically, like in the Kalman filter (Kalman 1960). If the evolution equation is nonlinear and/ or the probability density functions of the state and observation are not Gaussian, only an approximation of the integral in the Eq. A.3 would be available (Doucet 1997).

In case of a general nonlinear problem, one has to make recourse to approximate solutions, either by approximating the nonlinearity via successive linearization of the evolution equation (Corigliano and Mariani 2004) like in the extended Kalman filter, or via discrete approximate representation of the probability density function of the state vector (Mariani and Ghisi 2007; Doucet and Johansen 2009). The first remedy has broadly been applied to weakly nonlinear dynamic systems, and the required conditions for its stability have been extensively investigated (Ljung 1979). However, in some cases it is difficult, or even impossible to linearize the evolution equation. Moreover of severe nonlinearities may prevent the EKF from obtaining proper performance; hence, a category of filters have been developed to numerically handle the integrals in the Eq. A.3 (Kitagawa 1996). The aforementioned methods can be divided into two main categories: filters that are based on a Gaussian approximation of the probability density function of the uncertainties in the state, such as the sigma-point Kalman filter (Julier et al. 2000) and the Gaussian sum filter (Ito and Xiong 2000); filters that assume general form for the probability density function of the uncertainties in the state, such as the particle filter (Ristic et al. 2004) and the Rao-Blackwellized particle filter (Grisetti et al. 2007). Both the aforementioned categories of filters have known problems: the first class fail to provide accurate estimations in case of severely nonlinear and non-Gaussian problems; the second category can enhance the results but also drastically increase computational costs. This has motivated the researchers to develop a synergy of both approaches improving the numerical treatment of the integration in the Eq. (A.3) through an enhancement of the random quadrature procedures used by the particle filters. This notion has led to development of unscented particle filter (Van Der Merwe 2000), Gaussian mixture particle filter (Kotecha and Djuric 2003) and extended Kalman particle filter (de Freitas et al. 2000). From what preceded, one can conclude that the performances of recursive Bayesian filters in terms of computational burden and accuracy of the estimates may vary dealing with different problems. Aforementioned fact substantiates an assessment and comparison of performances once dealing with a specific problem. The Chap. 2 of this book provides and extensive study on the applications of recursive Bayesian filters to a shear-type building.

References

Arulampalam MS, Maskell S, Gordon N, Clapp T (2002) A tutorial on particle filters for online nonlinear/non-Gaussian Bayesian tracking. IEEE Trans Sig Process 50:174–188

Bittanti S (2004) Identificazione dei modelli e sistemi adattivi. Pitagora, Milan

Cadini F, Zio E, Avram D (2009) Monte Carlo-based filtering for fatigue crack growth estimation. Probab Eng Mech 24:367–373

Corigliano A, Mariani S (2004) Parameter identification in explicit structural dynamics: performance of the extended Kalman filter. Comput Methods Appl Mech Eng 193:3807–3835

de Freitas JFG, Niranjan MA, Gee AH, Doucet A (2000) Sequential Monte Carlo methods to train neural network models. Neural Comput 12:955–993

Doucet A (1997) Monte Carlo Methods for Bayesian estimation of hidden Markov models: application to radiation signals Unpublished doctoral dissertation, University Paris-Sud, Orsay, France

Doucet A, de Freitas N, Gordon N (2001) Sequential Monte Carlo methods in practice. Springer, New York

Doucet A, Johansen AM (2009) A tutorial on particle filtering and smoothing: fifteen years later. Handbook of nonlinear filtering 12:656–704

Grisetti G, Stachniss C, Burgard W (2007) Improved techniques for grid mapping with rao-blackwellized particle filters. IEEE Trans Robot 23:34–46

Haykin S (2001) Kalman filtering and neural networks. Wiley, New York

Ito K, Xiong K (2000) Gaussian filters for nonlinear filtering problems. IEEE Trans Autom Control 45:910–927

Julier SJ, Uhlmann JK (1997) New extension of the Kalman filter to nonlinear systems. In: Proceedings of SPIE—the international society for optical engineering, pp 182–193.

Julier S, Uhlmann J, Durrant-Whyte HF (2000) A new method for the nonlinear transformation of means and covariances in filters and estimators. IEEE Trans Autom Control 45:477–482

Kalman RE (1960) A new approach to linear filtering and prediction problems. J Basic Eng 82:35–45

Kitagawa G (1996) Monte Carlo filter and smoother for non-Gaussian nonlinear state space models. J Comput Graphical Stat 5:1–25

Kotecha JH, Djuric PM (2003) Gaussian particle filtering. IEEE Trans Sig Process 51:2592–2601

Ljung L (1979) Asymptotic behavior of the extended Kalman filter as a parameter estimator for linear systems. IEEE Trans Autom Control 24:36–50

Ljung L (1999) System identification. Theory for the user, 2nd edn. Prentice Hall, Englewood Cliffs

Mariani S, Ghisi A (2007) Unscented Kalman filtering for nonlinear structural dynamics. Nonlinear Dyn 49:131–150

Ristic B, Arulampalm S, Gordon NJ (2004) Beyond the Kalman filter: particle filters for tracking applications. Artech House Publishers, London

Van Der Merwe R, Doucet A, de Freitas N, Wan E (2000) The unscented particle filter, In: Dietterich TG, Leen TK, Tresp V (eds): advances in Neural Information Processing Systems (NIPS13), pp 584–590

Index

S. Eftekhar Azam, *Online Damage Detection in Structural Systems,*
PoliMI SpringerBriefs, DOI: 10.1007/978-3-319-02559-9,
© The Author(s) 2014